Retooling THE U.S. Housing Industry

Retooling THE U.S. Housing Industry

How It Got Here, Why It's Broken, and How to Fix It

Sam Rashkin

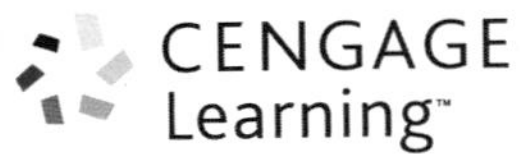

Retooling the U.S. Housing Industry
How It Got Here, Why It's Broken, and How to Fix It
Sam Rashkin

Vice President, Editorial: Dave Garza

Director of Learning Solutions: Sandy Clark

Senior Acquisitions Editor: James DeVoe

Managing Editor: Larry Main

Product Manager: Mary Clyne

Editorial Assistant: Cris Savino

Vice President, Marketing: Jennifer Baker

Marketing Director: Deborah Yarnell

Marketing Manager: Kathryn Hall

Production Director: Wendy Troeger

Production Manager: Mark Bernard

Content Project Manager: Mike Tubbert

Art Director: Casey Kirchmayer

Cover: ©DNY59/iStockphoto_1176658

Library of Congress Control Number: 2011927621

ISBN-13: 978-1-111-31382-1

ISBN-10: 1-111-31382-2

Delmar
5 Maxwell Drive
Clifton Park, NY 12065-2919
USA

Cengage Learning is a leading provider of customized learning solutions with office locations around the globe, including Singapore, the United Kingdom, Australia, Mexico, Brazil, and Japan. Locate your local office at:
international.cengage.com/region

Cengage Learning products are represented in Canada by Nelson Education, Ltd.

To learn more about Delmar, visit **www.cengage.com/delmar**

Purchase any of our products at your local college store or at our preferred online store **www.cengagebrain.com**

Printed in the United States of America
1 2 3 4 5 6 7 14 13 12 11

Table of Contents

Preface

Housing: It's Personal

This book is more than an opportunity to present a story about one of the nation's largest and most important industries. It's more than the story of how this industry began, grew through numerous technological developments and business cycles, and arrived today at what is likely to be its most critical juncture in history, facing the opportunity often presented in a crisis to completely reinvent itself. It's more than this story because it's personal.

It's Personal Because I Love Housing

At 6 years old I had a favorite toy called "Panel and Girder." It was an ingenious assembly toy in which plastic pegs resembling steel columns and beams could be assembled like a structural frame on a pegboard base and finished with plastic panels that looked like curtain walls. I spent endless hours creating my own skyscrapers. By age 12, I was sketching and designing homes and would build a model for my best friend's parents as if they were architectural clients. When I was finally allowed to select an elective course in middle school, I immediately chose mechanical drafting. It remained my selected elective each year through 12th grade. I went on to college and completed my bachelor's degree in architecture.

An elective course on solar energy and energy efficiency established a complementary passion for sustainable buildings long before this field became fashionable. After the prerequisite three years working with a licensed architect, I passed my licensing exams. Soon after, I hung out my shingle and started my own architectural practice, excited to specialize in energy-efficient housing.

But in the late 1970s, the economic gods intervened. On the heels of two oil supply disruptions, mortgage interest rates were soaring to 20 percent and beyond, effectively killing interest in residential projects. Forced to diversify my practice with less appealing commercial and government projects, I decided to pursue a career shift that would let me focus on both housing and energy efficiency. First, I secured a list of the nation's 100 largest builders and sent each a letter prospecting for interest to develop an energy-efficient product line. My efforts resulted in a full-day interview with one builder and a psychological profile test with another, but no job in the housing industry. Eventually, I secured a position managing energy-efficiency and renewable energy programs for the State of California. I also developed a part-time architectural practice and accepted only energy-efficient residential projects. Admittedly, many clients did not share my passion for energy efficiency, so I would secretly integrate it into the design process.

This career path led to a call one night at home in California inviting me to lead a national program for the U.S. Environmental Protection Agency (EPA) called ENERGY STAR® for Homes. It was 1994, and ENERGY STAR was far from the widely recognized product label it is today. However, the opportunity to work voluntarily with the U.S. housing industry to construct energy-efficient new homes was a dream job. Sixteen years later, there are more than 1 million ENERGY STAR labeled homes with more than 20 percent market penetration nationwide.

This long journey through many facets of the housing industry has provided me with a unique perspective and experience and has inspired this book. I have spent a career trying to make housing better; it's personal.

It's Personal Because I Live in a House

My wife and I have bought three homes so far. In every case, these homes have served us well, providing both enjoyment and financial appreciation. The latest move to the Washington, D.C. area led us to a beautiful home in Virginia on a south-facing lot surrounded with mature trees in a great neighborhood with natural lush mid-Atlantic landscape, wonderful open spaces, and great public schools. We looked long and hard for this kind of community, unwilling to live in a traditional suburban tract home development. However, we knew a major renovation would be needed, just as we had done on our prior two homes. And this was a relatively new home, just over 10 years old.

Wearing my homeowner's hat, each of my three homes inspired the same key question: Why do mainstream builders across the country typically provide low- or bottom-grade finishes (e.g., trim, flooring, cabinets, hardware, electrical outlets and switches, appliances, plumbing fixtures, and especially light fixtures) and minimum energy code features (e.g., insulation, windows, air sealing, heating and cooling equipment)? In other words, how did minimum first cost become the dominant driving force for the housing industry?

This is in direct conflict with my personal observations that homeowners don't appear to routinely choose the cheapest products. After all, homes represent the largest asset for most households and have significance far beyond basic shelter. It is often said that builders don't construct homes—they fulfill dreams. If that is true, my wife and I have found ourselves altering the low-cost dream builders seem to have had in mind for home buyers as we routinely removed and replaced cheap products used throughout all three homes we purchased. I live in homes builders build, and that makes the housing industry personal.

It's Personal Because Housing Is My Career

When I arrived in Washington, D.C. to work on ENERGY STAR for Homes, the division director made it clear my job security depended on substantial results transforming the housing industry to greater energy efficiency. This was not your typical government job, but it proved to be a great career decision. In the process, I've had the opportunity to train and engage thousands of housing industry stakeholders and promote advanced building practices. As a result, I have become good friends and acquaintances with builders, subcontractors, manufacturers, suppliers, sales and marketing professionals, energy consultants, utility program managers, and other government housing program managers all across the country. As we work through this unprecedented downturn in housing, many of them are affected emotionally and financially. Housing is very personal.

How to Use This Book

Retooling the U.S. Housing Industry is a must-read resource for anyone interested in housing, how this major U.S. industry got so far off base, and how it can get back on track. This text uses the recent housing market crisis as a learning moment for improving all aspects of delivering new homes in the United States. Students in construction management and architecture programs, along with practicing professionals in housing and all related activities, will benefit from this blueprint for rebuilding the entire industry.

Organization

Retooling the U.S. Housing Industry begins with a review of the current housing crisis and new trends that make it a critical time for significant changes. Chapters 2 through 6 are devoted to each of five critical components identified for the housing industry:

- **Sustainable Land Development**
- **Good Housing Design**
- **High-Performance Homes**
- **Quality Home Construction**
- **Effective Home Sales**

These chapters present a consistent analytical framework for each component: a review of key goals and how they are achieved, an overview of core principles, a review of how the industry arrived at current practices, an evaluation that reveals significant shortcomings with these practices relative to the key goals, and recommendations for resolving the shortcomings. The book concludes with a summary of recommendations for retooling the housing industry and how they can provide a profound marketing advantage for builders who apply them.

Features and Benefits

Retooling the U.S. Housing Industry is a comprehensive assessment of one of the nation's most important industries. Along the way, readers will discover the following:

- **Simple planning techniques that can lock in better communities for centuries**
- **What constitutes good design**

- **Why homes work and fail when it comes to providing affordable, comfortable, healthy, and durable living environments**
- **Practices and technologies that can substantially improve construction quality but are largely ignored**
- **How tried-and-true sales processes are completely absent in home sales today but essential for selling retooled homes**

By following these retooling recommendations, the housing industry can radically improve the quality of their product, dramatically increase customer satisfaction, and substantially reduce risks associated with litigation and high service center costs. All readers interested in joining the industry, supporting the industry, or staying relevant in the industry will significantly benefit from the opportunities for improvement presented here. Moreover, the analysis is presented in a uniquely understandable and efficient format including:

- **Insights from a long-time industry insider and innovator with key experience in energy-efficient design**
- **Clearly defined problems and concrete steps for addressing them**
- **Clean, clear graphics outlining the problem-solving approach and sound arguments supporting the analysis**
- **Conclusions for each chapter that summarize key findings and recommendations**

Resources for Instructors

This text is accompanied by an Instructor Resources CD to provide the following classroom support for teachers:

- **Computerized test banks in ExamView® software**
- **PowerPoint® presentations**
- **An Image Gallery including all text figures**

Inspiration

Writing is a creative endeavor. Creating anything is about 90 percent inspiration; the rest is process. A good idea for a book is likely to go nowhere without the support, confidence, and encouragement of a loved one. Enter Suzanne, my life partner. She has never stopped believing in my ideas, my abilities, and me for more than 30 years. It all begins and ends with her. Monica and Amanda inspire me every day because they bring a sense of joy and connection truly unique to children. My dream team at ENERGY STAR for new homes (Jon, Zak, Brian, Glenn, Ga-Young, and David) have inspired me with the most amazing collaboration a person could ever hope to experience in a career. I have the privilege to call a large group of builders across the country friends. They fulfill dreams every day. What could be more inspiring than that? My building science mafia buddies (Joe, Joe, John, Mark, Brad, and Gord, to name a few) inspire me with their amazing message, dedication, charisma, and passion. I can only hope this book lives up to their incredibly high standards. This is my first book, and I am tremendously indebted to my publisher who took on a new author. More important, they nurtured the development process with concise, cogent input that substantially improved this book. What else can you ask from a publisher?

About the Author

Sam Rashkin has managed ENERGY STAR for Homes since its inception in 1996. Under his leadership, ENERGY STAR for Homes has grown exponentially to more than 9,000 builder partners and more than 1 million labeled homes. He is also overseeing an advanced technology program for elite builders as part of the ENERGY STAR program.

Mr. Rashkin earned his bachelor's degree in architecture at Syracuse University. He completed Masters of Urban Planning studies at New York University and is a registered architect in California and New York. During his 20-plus years as a licensed architect, he specialized in energy-efficient design and completed more than 100 residential projects. He has served on national steering committees for USGBC's LEED for Homes, NAHB's Green Builder Guidelines, and the U.S. EPA's WaterSense label. He has also served on the development team for the U.S. EPA's Indoor airPlus label. Mr. Rashkin has prepared hundreds of articles, technical papers, reports, and seminars and has contributed to numerous books and Web sites covering a wide range of energy subjects.

Retooling THE U.S. Housing Industry

1

Why It's Time to Retool: A Learning Moment Inspired by Crisis

In 1997 a loud and clear signal portended the imminent collapse of the housing industry—but no one heard it. The roar of wild profits being made selling, financing, and flipping homes in the midst of an unprecedented boom period drowned out the signal. Because the warning came from the weak sister in the world of housing—manufactured homes—no one paid attention.

Manufactured homes are built in a factory to a national standard promulgated by the Department of Housing and Urban Development (HUD). This national code helps the industry avoid potentially prohibitive costs that would otherwise be necessary to accommodate the myriad building code requirements across the large geographic areas served by each plant. These homes are officially called HUD-code homes, and HUD's involvement is driven by its mission to serve the housing needs of low-income families.

Manufactured housing grew to more than 180,000 homes per year by the 1990s and came to represent the largest segment of affordable market-based housing. Under pressure from the investment community to keep increasing sales, the industry resorted to a typical growth strategy often employed by corporate America: creative financing. Specifically, the "chattel loan" became widely available in the 1990s, offering manufactured housing retailers a nonrecourse consumer financing option similar to car loans. The chattel loan was tied only to the asset (the home) and not to the property. It facilitated increased sales by enabling buyers to purchase homes often with little or no down payment, limited employment history, and weak credit. However, interest rates were much higher than a typical mortgage to make up for the substantially greater risk. With help from a strong economy and the chattel loans, annual production of manufactured homes nearly doubled to more than 335,000 in 1997. Then came the warning shot.

© iStockphoto.com/MCCCAIG.

© iStockphoto.com/Purdue9394.

Suddenly, and predictably in hindsight, chattel loans began to experience massive defaults. The frail fiduciary underpinnings buckled as a growing number of homeowners with limited financial resources failed to make their payments. One contributing factor was relatively poor energy efficiency common to manufactured homes. This left many homeowners with very high and unexpected monthly utility bills, often comparable to their monthly chattel loan payments. As homeowners defaulted, these homes were dumped into the used manufactured homes market at substantially discounted prices. It is not surprising that almost all chattel loan lenders soon pulled out of the market.

The perfect storm was complete. Manufacturing plants were churning at unsustainable production levels while the market of qualified buyers was substantially cut without chattel loan financing. This problem was compounded by additional competition from an increasing number of default homes at fire-sale prices. The HUD-code industry went into a sudden and deep swoon in 1997 with production dropping more than 70 percent to about 90,000 homes. By 2009 only about 40,000 manufactured homes were produced by an industry that had produced nearly nine times that volume in recent years. Over half the manufacturers went out of business or were acquired by surviving companies. Some experts are concerned that the entire industry may disappear.

But the world of site-built housing didn't even blink at these extraordinary events. Good times were too intoxicating and profits continued to roll, at least for another decade. In 2007, when the exact same scenario played out in site-built housing, somehow the world was caught by surprise.

History Repeats Itself, Go Figure

Site-built housing enjoyed an unprecedented boom in sales that began in the early 1990s. Along with strong sales came soaring land and material costs. In many cases builders were happy just to get building materials even at substantially inflated prices as global competition created scarcity issues for some products. For instance, China's exponential economic growth and ensuing construction boom were considered leading factors behind a worldwide shortage for concrete and wood framing.

Just as with manufactured housing, investors were putting pressure on conventional single-family home builders to keep increasing sales. Company acquisitions offered a temporary solution and led to an industry consolidation. In 1997 the top 10 largest builders captured 11 percent of the single-family for-sale market, but their market share had almost doubled to 20 percent by 2002.[1] However, acquisitions could only get builders so far. Cue the typical business response to attracting more customers: creative financing. In the world of conventional real estate, subprime loans took creative financing to historic new levels of risk.

Lenders started with the basics, providing mortgages to buyers who could not provide any down payment, documentation of adequate income, or good credit scores. Then the lending industry got even more creative with options such as supersized loans, incredibly low initial teaser rates to mask the real carrying costs lying ahead, and extended loan terms up to 40 years. Buyers were willing to stretch far beyond normal income limit constraints based on claims that the huge appreciation rates in past years would continue indefinitely. These claims were often voiced by prominent economists.[2] If this sounds a bit like a Ponzi scheme, it should. Those who got in early made money, and those who came in late were likely to lose everything. Only in this case, buyers often had no or minimal "skin in the game" because they did not contribute any down payment, leaving investors on the short end.

There is plenty of blame to go around. Federal regulators, major lending institutions, appraisal organizations, and many others in positions of authority and leadership who should have known better were not providing any adult supervision. In 2007, a decade after the collapse of manufactured housing, site-built housing followed suit. Only this collapse was even more impressive, taking the rest of the economy along with it.

A Housing Crisis of Historic Proportions

The U.S. housing industry has always been cyclical, with large peaks and valleys. However, the current housing crisis is probably going to go on record as one of the most severe and protracted. U.S. Census data indicate total production in 2009 had fallen from a peak of more than 2 million units in 2005 to about 570,000 units in 2009 (Figure 1.1). The percentage decline across many of the nation's hottest housing markets is unprecedented:[3]

- **Arizona was down nearly 80 percent, from its peak of nearly 60,000 homes to approximately 14,000 homes.**
- **California was down almost 85 percent, from its peak of more than 202,000 homes to just under 34,000 homes.**

---

[1] Elaine F. Frey, "Building Industry Consolidation,". *Builder Magazine*, August 2003.

[2] David Lereah, *Are You Missing the Real Estate Boom?* (New York: Crown Publishing Group, 2005).

[3] "Housing Units Authorized by Building Permits: Table 2 – United States Region, Division and State (Unadjusted Data)," U.S. Census Bureau, December 21, 2010, www.census.gov/const/www/C40/tablel2.html

Figure 1.1: Housing Units Authorized by Building Permit*.

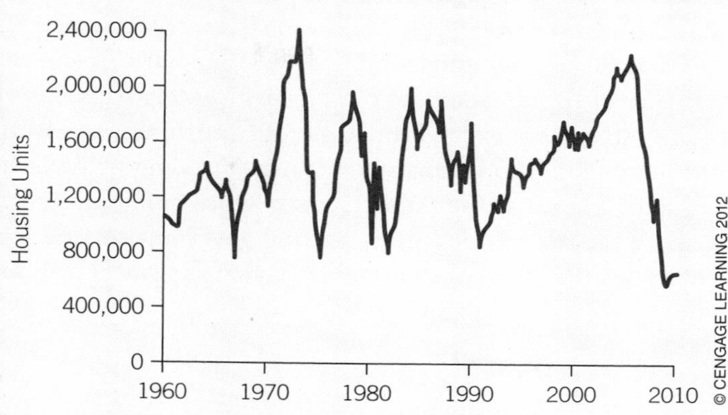

Source: "Housing Units Authorized by Building Permits: Table 2 – United States Region, Division and State (Unadjusted Data)," U.S. Census Bureau, December 21, 2010, www.census.gov/const/www/C40/tablel2.html

* New Private Housing Units Authorized by Building Permits (PERMIT)," Economic Research, Federal Reserve Bank of St. Louis, December 21, 2010, http://research.stlousisfed.org/fred2/series/PERMIT

- **Florida was down over 85 percent, from its peak of just over 285,000 homes to less than 36,000 homes.**
- **Georgia was down nearly 85 percent, from its peak of nearly 105,000 homes to just over 17,000 homes.**
- **Nevada was down over 85 percent, from its peak of just over 47,000 homes to less than 7,000 homes.**

The housing crisis has had a profound impact on the nation's entire economy because it remains one of the few significant products wholly made in the United States. At its peak in 2006, housing accounted for 6.2 percent of gross domestic product (GDP). In the first quarter of 2010, housing accounted for only 2.4 percent of GDP. Since the recession began at the end of 2007, 1.9 million residential and commercial construction jobs have been lost, with additional losses at real estate offices, mortgage processors, and other affected transaction industries. Construction also feeds into manufacturing, accounting for some 10 percent of U.S. manufacturing orders for building materials and supplies. An additional 2 percent of manufacturing orders are for household appliances and furniture. All of these industries are depressed by the housing industry slump.[4] This crisis has been extremely difficult for many regional economies vested in housing industry materials and products, especially in the midwest.

[4] Justin Lahart, "Building Supplies Drag Retail Sales," *Wall Street Journal,* July 21, 2010.

The ripple effects spread outside the industry as well. It began with the reduced need for jobs and services that support housing industry businesses and workers (e.g., accounting, restaurants, cleaning, automobiles). The National Association of Home Builders (NAHB) analysis shows that the average single-family new home buyer spends on average $7,400 more than a similar homeowner who does not move, including $4,900 in the first year after purchase.[5] It is likely that these expenditures are for furnishings, accessories, and improvements. That added more than $10 billion of consumer spending by households that moved into newly constructed homes at the industry peak. Once homes are on the ground, they continually contribute tax revenue resources to local governments and stimulate economic activity in communities. The 75 percent drop in new housing starts truly affects the nation's economy.

The New Normal

If you are a home builder, you're not in Kansas anymore (with full apologies to those builders who actually are in Kansas). The following sections address some of the critical changes affecting the business case for constructing new homes.

Smaller Universe of Qualified Buyers

More rigorous credit guidelines are here to stay, starting with responsible mortgage practices. Now you actually have to have a down payment, documentation of a job with adequate income, and a reasonable credit score before you can buy a home. Excuse the sarcasm, but it was extremely disturbing to watch lending practices get so outrageously irresponsible. Similarly, banks are completely revamping consumer credit policies because looming bad credit debt is the next toxic asset that can further drag down the economy. As a result, credit card eligibility requirements are becoming more rigorous and credit limits are falling. Along with tighter credit, as the economy continues to struggle, fewer people have the job stability needed to qualify for a mortgage. And many people who have secure jobs are much less confident that their salaries will increase as they have in the past. This will affect their ability to purchase homes because buyers often counted on income growth to manage the financial obligations associated with homeownership. It all adds up to dramatically fewer people who are qualified to buy homes.

More Competition from Low-Priced Used Homes

A recent Zillow.com survey reports that 8 percent of homeowners, or about 10 million Americans, are "very likely" to lose their homes, combined with the foreclosures already in the pipeline.[6] More than 1.2 million U.S. properties in some stage of foreclosure sold to third

---

[5] Natalia Siniavskaia, "Spending Patterns of Home Buyers," NAHB courtesy of HousingEconomics.com, December 4, 2008, http://www.nahb.org/generic.aspx?sectionID=734&generiContentID=106491&channelID=311

[6] Aaron Task, "Housing's Big 'Shadow': Up to 10M More Homes Could Be for Sale, Zillow.com Says," *Yahoo Finance.com,* March 24, 2010, http://finance.yahoo.com/tech-ticker/housing%27s-big-%22shadow%22-up-to-10m-more-homes-could-be-for-sale-zillow.com-says-448362.html?tickers=XHB,PHM,DHI,LEN,TOL,HD,LOW&sec=topStories&pos=9&asset=&ccode=

parties in 2009, with the average sales price of those properties 25 percent below the average sales price of properties not in the foreclosure process.[7] Moreover, 5 million to 7 million properties are potentially eligible for foreclosure but have not yet been repossessed and put up for sale.[8] In addition, nearly one in four U.S. homeowners with a mortgage, or about 11.3 million households, owed more on his or her mortgage than the home was worth by the end of 2009.[9] These mortgages are referred to as "under water." All of these developments are a sobering reality, suggesting that we are in for a long period of downward price pressures from distressed sales. In numerous markets, builders are concerned that they cannot afford to construct new homes at these reduced price points.

Urban Centers Becoming More Geographically Desirable

There are approximately 76 million baby boomers (born between 1946 and 1964) and 72 million Gen Y youth (born between 1977 and 1994).[10] Many studies suggest that both of these groups are looking for smaller homes in urban centers.[11] Baby boomers are looking for less maintenance and closer proximity to work and cultural amenities. Gen Y youth ready to live on their own are looking for simple lifestyles without the burden of home maintenance and good with access to nightlife and activities. These trends will dramatically reduce the demand for conventional suburban developments, at least until the Gen Y demographic start to have their own families.

Increasing Perception That Homeownership Is No Longer Compelling

As soldiers returned home after World War II, new policies and mortgage programs fueled the rapid development of suburbs across the country. As a result, homeownership increased from a low of about 44 percent before World War II in 1940 quickly to 62 percent by 1960 and then slowly to a peak of about 67 percent between 2000 and 2007.[12] The current economic crisis has substantially curtailed this dream, and the homeownership rate has dropped for the past five years. As unemployment became a critical issue across the nation, many people felt they needed more geographic flexibility to access fewer jobs. As a result, being locked into a single location with home ownership became less appealing due to concerns of a long and costly

---

[7] "RealtyTrac: Foreclosed Homes 31% of Q1 US Residential Sales," Need to Know News, Wednesday, June 30, 2010, www.automatedtrader.net/real-time-news/46602/realtytrac-foreclosed-homes-31of-q1-us-residential-sales

[8] Stephanie Armour, "Underwater Mortgages Drain Equity, Dampen Retirement," *USA Today*, March 26, 2010, www.usatoday.com/money/economy/housing/2010-03-24-1Aunderwater25_CV_N.htm

[9] Renae Merle and Dina ElBoghady, "U.S. Launches Wide-Ranging Plan to Steady Housing Market," *Washington Post*, Thursday, March 5, 2009, http://www.washingtonpost.com/wp-dyn/cotent/article/2009/03/04/AR2009030400911.html

[10] Frederick G. Crane, Roger A. Kerin, Steven W. Hartley, William Rudelius, *'Principles of Modern Marketing, 7th Cdn Edition,'* (New York: McGraw-Hill, 2008), 69–70.

[11] Katy Tomasulo, "Boomers, Gen Y, Immigrants to Shape Housing Demographics," *EcoHome*, November 6, 2009, http://www.ecohomemagazine.com/news/2009/11/boomers-gen-y-immigrants-to-shape-housing-demographics.aspx?printerfriendly=true

[12] Linda A Jacobsen, Mark Mather, "U.S. Economic and Social Trends Since 2000," Population Reference Bureau Program Bulletin, Vol. 65, No. 1, February 2010, p. 7.

sales process to break free (personal observations from travel across the country). This applies to people who are already out of work, those on a career path where layoffs are a concern, as well as those gainfully employed but seeking to keep their options open for better jobs.

Most significant, the hyperinflated home appreciation Ponzi scheme is over. As a result, real estate is no longer the "golden child" for asset growth. As consumers lose confidence that homeownership will realize meaningful appreciation, the costs and burden associated with maintenance weigh more heavily in the decision to purchase a home. This includes costs for both routine upkeep and large repairs that come up periodically (e.g., replacing the roof, furnace, windows, or siding). Renting or the unthinkable, moving in with parents, have become much more viable options for many. In 2006 the American Dream collided with reality. Now, many homeowners and homeowner wannabes understand that "not everybody can afford or should own a home."[13]

A Protracted Economic Recovery

I was invited to speak right after a prominent housing industry economist at a builders conference in the spring of 2007. At the front end of the economic crisis, this economist boldly predicted that the down cycle would last only into the middle of the following year and then the industry would experience a quick recovery. I immediately recalled a quote from a business professor's keynote speech at my daughter's recent college graduation ceremony, "Without data, you're just another jerk with an opinion." I cringed at this well-intended inspirational advice believing it was misguided for young graduates heading into the real world. I refer readers to Malcolm Gladwell's book Blink for an alternative perspective that explores how rapid cognition based on experience and instinct can often prove more critical than research and data.[14] I was just about to speak after a credible expert who expressed a forecast based on extensive data that I thought was utter nonsense.

I proceeded to offer an admittedly nonexpert alternate economic forecast to the audience based on experience and instinct. I'll summarize: The "jig is up" on the creative financing–driven economy. Tighter credit and increased savings are inevitable and here to stay. Anything less is a house of cards. As a result of less credit and more savings, we will be buying less "stuff." All stuff, not just houses. Less furniture, jewelry, cars, vacations, and appliances, and the list goes on. If we are buying less stuff, industries across the country will have to scale back and fewer people will be needed to make stuff. If we are making substantially less stuff, fewer people will be working. All the related industries that service and supply the industries and employees experiencing cutbacks will have to scale back as well. This is the ripple effect. I recommended that folks hang on for a long bumpy ride; this housing slump has a long way to go.

To achieve a permanent source of new jobs based on responsible consumer behavior, profound challenges lie ahead to develop new value-based technologies and services (e.g., demand tied to a compelling value proposition as opposed to a financing scheme). I suspect

---

[13] Jonathan D. Miller, *Emerging Trends in Real Estate®* 2010 (Washington, DC: Urban Land Institute and PriceWaterhouseCoopers LLP, October 2009), 54.

[14] Malcolm Gladwell, *Blink*, Little, (New York: Brown and Company Time Warner Book Group, 2005).

(and hope) that we will get there sooner rather than later, with many experts touting new energy, bioengineering, and nanotechnologies as likely sources of high-paying new jobs. But developing these new industries, including manufacturing and distribution infrastructure, takes time . . . a lot of time. This is the reason I believe the economic experts who have got it right are those predicting a long, protracted economic recovery.

The Learning Moment Has Arrived

This new normal should be sobering to anyone involved in the housing industry. But it starts with builders who have to figure out how to:

- **Compete for a much smaller universe of home buyers**
- **Compete more effectively against lower priced used homes . . . often priced lower than their costs to build**
- **Attract home buyers to locations away from geographically preferred urban centers**
- **Overcome the perception that homeownership is no longer compelling**
- **Sustain sales during a long, protracted economic recovery process**

This is a far cry from the "build it and they will come" good times. More square footage, 10-foot ceilings, and granite countertops will no longer "get 'er done." In effect, the housing industry has to reinvent its entire value proposition. Those who figure it out will succeed; those who do not are much more likely to fail. A crisis is indeed a learning moment. In that spirit, I have broken down the housing industry into five key components: Sustainable Land Development, Good Housing Design, High-Performance Homes, Quality Home Construction, and Effective Home Sales. These components, along with their most critical goals and how they are achieved, are illustrated in Figure 1.2.

Here's the Spoiler: Every Component Is Profoundly Broken

Over the past 50 years we have been developing communities that compromise livability goals with isolated neighbors, automobile dominance, lack of open spaces, minimum-cost landscaping and exterior appointments, and development shortcuts that wipe out natural settings and ignore natural comfort. Homes have become bigger but not better, with less design attention to trim and detail, wasted space, and little integration of mechanical, structural, electrical, plumbing, and information technology systems. Living in homes is compromised by poor performance resulting in less comfort, poor indoor air quality, and durability problems. Moreover, these performance goals could all be improved for less ownership cost because the added monthly mortgage cost attributed to home performance improvements is easily offset by monthly utility savings. Instead, under the pressure of lowest "first-cost" construction, the housing industry fails to consistently employ the latest technologies, building systems, and quality assurance techniques that both serve builders' bottom-line business objectives and deliver a better product for

Figure 1.2 Key Components of the U.S. Housing Industry.

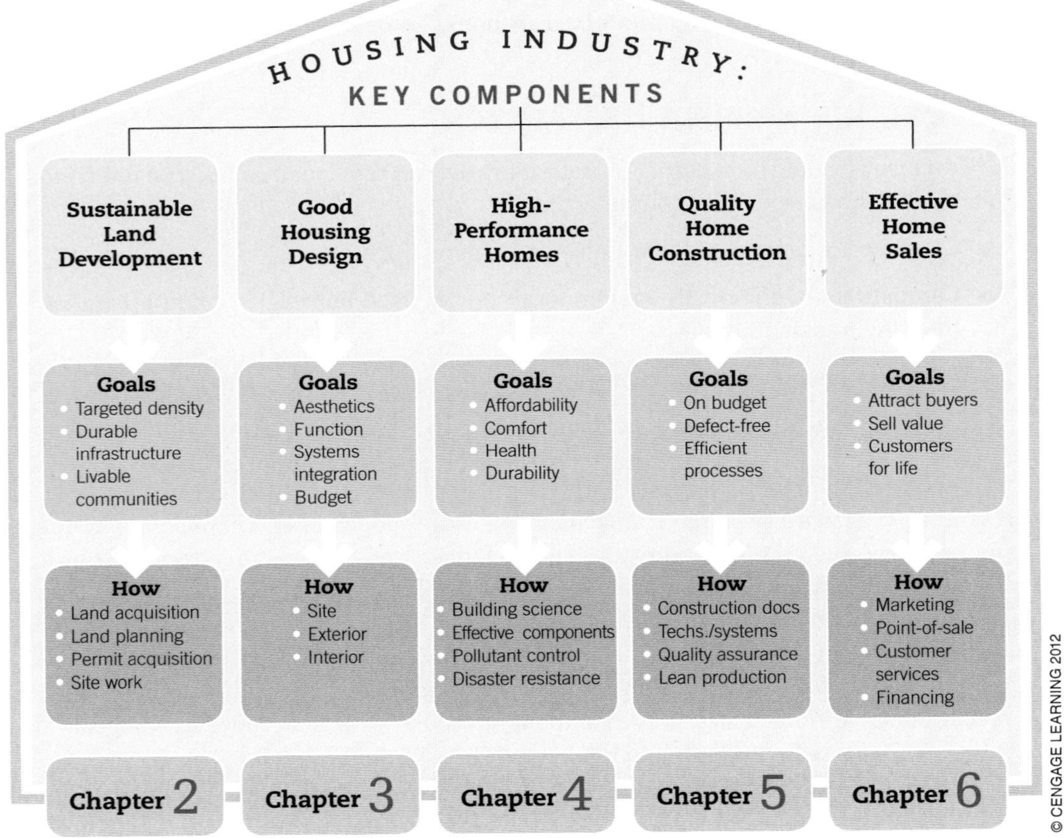

home buyers. And the sales infrastructure for the housing industry is ill prepared to sell quality and performance benefits you cannot see. This fuels a well-established industry misconception that home buyers will not pay for better homes. You cannot expect consumers to value quality and performance improvements if you do not sell them effectively.

Each of the next five chapters evaluates a key component for the nation's mainstream single-family housing industry and includes the following:

- **A quick summary of the process, goals, and how goals are achieved**
- **A description of what is entailed integrating the goals into the housing industry**
- **A review of developments that led to current industry practices**

- **A comparison of these practices with critical goals revealing why they are broken**
- **A set of recommendations for how to fix what is broken**

The last chapter compiles all of the recommendations into a comprehensive summary of the "big" ideas and examines how together they provide a compelling value proposition for new home buyers.

Chapter 2: Sustainable Land Development. This is where the new home process begins. Raw land is identified, secured, planned, permitted, and improved so that builders can construct homes on individual lots. Sometimes the builder and developer roles are combined, and sometimes they are split. Either way, the use patterns and features of each property often are locked in for one or more centuries. A sustainable development process addresses the profound effects of every decision regarding street configurations, home orientation, open space, and preservation of natural landscaping and drainage patterns. The choice is ***Livable Communities or Compounds.***

Chapter 3: Good Housing Design. One of the best definitions I have heard for "design" comes from a mentor early in my career: design is an iterative process toward an optimum solution. In the case of housing, critical design factors include the following:

- **Interior and exterior aesthetics**
- **Functional spaces and layout**
- **Systems integration**
- **Available budget**

This process will be substantially influenced by a major shift in design preferences toward smaller homes by the two largest demographic groups: baby boomers and Gen Yers. New home designs will need to consider more effective use of space, quality features, and new technologies along with ignored facets of design such as lighting, built-in cabinets, and natural comfort. This is critical because ***Design Trumps Everything.***

Chapter 4: High-Performance Homes. How homes perform would not make the list of key business components for most industry observers. However, the concept of a "high-performance home" has become increasingly recognized over the last few decades as an essential element for housing that works. This is because high-performance homes deliver:

- **Increased affordability based on full ownership costs (mortgage, utilities, and maintenance)**
- **Improved comfort through more even room-by-room temperatures, reduced noise, and improved moisture control**

- **Improved health by reducing moisture-related problems, dangerous chemicals, radon gases, dust, pollen, and pests**
- **Greater durability from better weather protection, less damaging UV sunlight, fewer moisture problems, and longer-lived equipment**

High-performance home goals are achieved by:

- **Applying basic building science principles to control air flow, thermal flow, and moisture flow**
- **Minimizing residual loads with energy-efficient components (e.g., heating and cooling equipment, lighting, appliances, and fans)**
- **Reducing pollutants with source control, dilution, and filtration**
- **Incorporating appropriate measures to fortify homes against region-specific disaster risks**

American home buyers will come to demand the compelling quality advantage of high-performance homes because they address ***Why Homes Work and Fail.***

Chapter 5: Quality Home Construction. Homes have historically been built with small incremental changes, adopting new technology and quality processes. In fact, the NAHB Research Center reports that it takes up to 25 years for the industry to implement new technologies. Yet, as mentioned, the largest demographic group of emerging new home buyers is Gen Yers, who embrace and thrive on new technology. The good news is that many new technologies and practices can dramatically improve quality and performance while reducing the risk of defect litigation and increasing customer satisfaction. It's time to ***Stop Protecting Old Technologies and Processes.***

Chapter 6: Effective Home Sales. Current sales practices in the housing industry emphasize appearance and cosmetic features. These are highly emotional visual experiences that home buyers readily understand. The challenge for effective home sales is to sell the invisible features and benefits of retooled homes: new homes in better developments, with superior designs, revolutionary improvements in home performance, substantially better quality construction, and advanced technologies and practices. Investments in the sales process are critical because if you don't tell the story, you will be giving valued improvements away for free, ***Better Is Not Good Enough.***

Chapter 7: Putting It All Together. I continue to travel across the country training thousands of builders, home energy raters, trade subcontractors, and state and utility program staff working on builder programs. As I do, I am encouraged by the progress made in the last 15 years integrating significant high-performance home improvements. However, this is only a good start. This time of crisis provides a unique opportunity to jump-shift the housing industry to a much higher gear and embrace the wholesale retooling so badly needed. All of the recommendations for fixing what is broken are compiled in this chapter to assist in this transformation. In addition, a theoretical brochure is presented that demonstrates the compelling value proposition available to builders who do apply the retooling recommendations in this book. If you build better communities, designs, performance, quality, and sales capabilities, they will come . . . again.

A Final Word

Retooling entails radical change for an industry highly resistant to change, but it is time. The housing industry has to do better, much better, to navigate the new normal. It cannot afford to ignore the substantial rewards that come with retooling: increased sales, reduced risks, and maximum profits. The concomitant outcomes for the rest of us will be increased jobs, less dependence on foreign energy sources, cleaner air, stronger communities, and higher-quality homes. Change is good for everyone, but it will not be easy.

Chapter 1 Review

So What's the Story?

The current housing crisis of historic proportions presents a unique opportunity to rethink how we build new homes, and that is the purpose of this book. The housing industry includes five major components: sustainable land development, good housing design, high-performance homes, quality home construction, and effective home sales. Each one of these key components is fundamentally broken. The good news is that each can be fixed, and in the process, make the housing industry dramatically stronger than it has ever been in the past. This "Why It's Time to Retool" story can be summarized as follows.

- **The housing industry was heading for a crisis, but somehow no one noticed the warning shot fired by the collapse of HUD-code housing a decade earlier.**
- **With the exact same set of conditions experienced by the HUD-code industry, history repeated itself for site-built housing, and everyone was surprised.**
- **This housing crisis is far deeper and will last longer than in years past, with significant impacts on the rest of the economy.**
- **As a result, the housing industry must adapt to a new normal business climate with fewer buyers, more competition, increasing preference for urban centers, and a protracted economic recovery.**
- **A crisis is the perfect time to embrace change, and change is needed in every major component of the housing industry.**

Retooling THE U.S. Housing Industry

2

Sustainable Land Development: Livable Communities or Compounds

Sustainable Land Development: Process, Goals, and How Goals Are Achieved

Process. *Residential land development requires an ability to envision the possibilities associated with a raw piece of land by looking beyond its existing natural condition or by enhancing an existing land use (e.g., add infill housing, convert polluted and abandoned industrial "brownfield" areas into desirable communities, or implement urban renewal projects). Developers can optimize profits by purchasing raw land early, before values rise, and by identifying undervalued existing developments that can be enhanced with a more market-appropriate product (density, price, mixed use), superior design, special amenities, and improved infrastructure. Once desired land is identified, land acquisition professionals have to choose the best option for locking in the land and negotiate favorable prices and conditions. Then the planning professionals complete designs and prepare final construction documents. Necessary permits called "entitlements" are secured and work proceeds installing land improvements and infrastructure. Thereafter, builders can begin construction on individual lots. This simple description belies a very complicated process with numerous minefields at*

Sustainable Land Development

Goals
- Targeted density
- Durable infrastructure
- Livable communities

How
- Land acquisition
- Land planning
- Permit acquisition
- Site work

© iStockphoto.com /jhorrocks.

© iStockphoto.com/dan_prat.

every step. This chapter focuses on how this basic land development process can incorporate natural and constructed amenities along with planning concepts and covenants that optimize property values on opening day and in perpetuity. This will lead to developments that are more sustainable from both a resource and a financial perspective.

Goals. *The first goal of sustainable land development is to provide adequate density (housing units per acre) to ensure the economic viability of the project. How this goal is set can be very controversial for any development claiming to be sustainable. Environmental groups continue to advocate for higher-density projects, often located close to urban centers. The next goal is to ensure that a durable infrastructure is in place that provides the full array of services needed to support each home. This includes water supply, storm and sewer systems, utility electric lines and piping for energy services, information networks, fire hydrants, streets, sidewalks, and roads. The critical final step is to ensure that land development results in livable communities. However, you will discover people have differing strong opinions on what is considered a livable community. In an attempt to find some common ground, I offer the following criteria:*

- **Ease of social connections**
- **Ability to foster a safe environment**
- **Preservation of natural attributes (e.g., trees, topography, drainage, vistas, and access to free solar energy)**
- **Generous open spaces and amenities (e.g., park areas, ponds, water features, trails, community gardens, town centers, recreational facilities)**
- **Regionally appropriate landscaping (e.g., new trees, turf, shrubs, year-round flower beds, and ground cover) and hardscaping (e.g., streetlights, landscape lighting, benches, fencing, trellises, and accent pavement)**

How. *Successfully reaching these goals begins with acquiring desirable raw land that does not have fatal flaws such as proximity to an undesirable land use, excessive slopes that are too difficult to develop, or low elevation in a flood-prone area. Early in a region's development, choice properties are often more readily available. However, as areas experience incremental growth, development opportunities often are limited to more marginal land that is farther from employment centers. Business strategies such as lower prices, more site amenities, and innovative design features can attract home buyers to these less desirable locations. Effective land planning clearly identifies site-specific criteria for livable communities and uses them to prepare a project design. Planning documents are then prepared based on the final design and are submitted to the planning department to secure the required permits. The actual site work must be executed efficiently and in conformance with the plans. This basic land development process is discussed in greater detail in the following section.*

What Is Sustainable Land Development?

Land development is often considered the most critical component of the housing industry because it drives most of the profits. This is based on the reality that gross margins from developing raw land far exceed gross margins for constructing homes. Although there is a lot of variation, one report suggests that land value increases 300 to 500 percent, on average, once it is fully developed and ready for home construction.[1] As a result, the housing industry's primary business goal has sometimes been cited as "flipping" land rather than constructing homes.[2] The following five-step land development process is based on guidance provided by an independent investment consultant, John Hanlin.[3] I have included additional requirements for sustainable land development in each step.

Step 1: Research. The objective is to identify desirable raw land in the path of growth that has not yet been entitled for development. A professional land developer or in-house land development manager for a large production builder will spend a great deal of time researching key factors that may affect potential profitability, including economic and related job growth and population trends relative to growth, age, and ethnic mix. In addition, research typically involves extensive networking at the regional and local level, including local government officials, business leaders, real estate brokers, builders, and environmental groups. Along the way, it is critical to research local government land-use maps to identify potential uses and adjoining property zoning which can have a significant impact

[1] JohnHanlin.com, "Land Development Process Step-By-Step," http://www.johnhanlin.com/Raw_Land_Development_101.html (accessed November 9, 2010).

[2] Gord Cooke, *Selling Energy Efficient Homes*, Presentation at Annual EEBA Conference, Minneapolis, MN, 2006.

[3] Ibid.

on land value. Sustainable development is integrated into this process by attempting to find sites with the most desirable trees and landscaping, opportunities for open spaces with special amenities, and the ability to maximize east-west-oriented streets.

Step 2: Land Acquisition. The objective is to buy desirable raw land at favorable prices based on the research results from Step 1. Numerous factors can affect the cost of land, including timing (e.g., prices increase after successful developments on adjoining land demonstrate economic viability of an area), economic developments (e.g., announcements of new manufacturing plants or business growth that will increase new local jobs), new amenities (e.g., parks, golf courses, mass transit projects, or protected forests), and public policy developments (e.g., smart growth constraints that increase demand for eligible raw land).

There are typically two options for securing land. The first is to purchase the raw land up front and thus own it outright. The second is to arrange for an "option to purchase agreement" that provides rights to the raw land at a set price by a set date. The former option can help secure property at the lowest cost, but it entails greater risk. The option to purchase is secured with a nonrefundable deposit and entails much lower risk (i.e., only the deposit). During the housing boom, builders that purchased land outright before prices inflated made impressive profits. Once the current industry crisis took hold, builders heavily vested in land secured after prices inflated were much more likely to experience bankruptcy problems than builders who secured land with "option" agreements. Even in less volatile times, a number of risks can affect the future value of land, including public policy changes such as setting urban growth boundaries (e.g., Portland, Oregon), determinations of endangered species on the raw land, and development moratoriums and high land development fees (e.g. one California builder informed me that development costs exceeded $80,000 per lot for permit fees, school fees, engineering fees, environmental impact studies, and workers' comp before they even put a shovel in the ground). Major economic and demographic shifts also may affect the value of raw land, but these factors should have been uncovered during the research phase. The major impact sustainable development should have on land acquisition is to instill a willingness to invest in sites with more natural assets.

Step 3: Architectural Plans. The objective of this step is to prepare architectural plans for the raw land. The developer has to define critical design attributes targeted for a specific property to begin the design process. It is the job of the land planner to incorporate these preferences into a workable design. Once a final design is approved, the architectural plans will lay out all site work, lots, easements, and details for improving the property. This is a critical time as the future for a specific property will be locked in potentially for hundreds of years by converting the "vision" to a set of legal documents including street configurations, lot sizes, maximum footprint area, and easements. When a developer also wants to specify one or more targeted architectural styles and exterior elements, "pattern books" are sometimes prepared. Builders are then obligated to comply with the specified architectural requirements. Sustainable development takes this process further by ensuring final plans preserve desirable trees and landscaping; maximizing east-west-oriented streets; specifying regionally appropriate and durable landscape and hardscape features; and including a generous amount of open space and amenities.

Step 4: Entitle Land. The objective of the entitlement process is to obtain the necessary building permits by the appropriate jurisdictions needed to commence site work. From a business perspective, this is a critical step in land development. By securing approvals to utilize raw land for a higher economic purpose, significant value is being created that can make land development very profitable. When developers choose to employ innovative design features such as narrow streets, natural drainage, and on-site sewage treatment concepts, substantial resistance is often met at local planning departments. Thus more time and costs are often required for innovative developments to navigate the entitlement process. One such project, Haymount in Fredericksburg, Virginia, was so innovative that the delayed approval process has taken the development from the height of the building boom to the current slump. Now it is uncertain whether this unique traditional neighborhood development will ever be completed, at least as originally envisioned. A hundred percent of nothing is still nothing; compromising with bold aspirations is sometimes critical. When land is entitled, sustainable developments incorporate covenants attached to property ownership that will ensure the development has the resources to maintain, nurture, and enhance the landscaping, hardscaping, and open spaces.

Step 5: Site Work. The objective of this final step is to complete all the improvements needed so construction can begin and generate income from the sale of lots. This entails site work by the developer to construct streets, amenities, and infrastructure so that each lot is builder ready. Typically the newly "entitled land" is sold by the lot or parcel to building contractors and other real estate developers. However, some builders develop the land as well as construct the homes. In either case, the process concludes when sites are ready for homes to be constructed and sold to home buyers. In sustainable development, this work conforms to all the plans that protect natural resources, enhance street orientation, and incorporate open space and amenities.

How Sustainable Land Development Got Here

Urban Development Dominated to the End of World War II

Until the 1950s, most development took place within the urban core with outlying land developed on a custom, lot-by-lot basis using custom architectural plans or designs from a plan book. Urban areas had the advantage of critical infrastructure required for roads, utilities, water supply, fire protection, schools, and access to jobs. Moreover, homeownership was not feasible for many Americans during this time because modern long-term financing concepts were not developed. Potential home buyers had to save the cash needed to purchase homes. As a result, urban areas were developed to serve the large demand for rental housing.

Suburban Development Dominated After World War II

Once World War II ended, government policy emphasized getting veterans into homes and back to work. This led to the GI bill and a home mortgage program that was intended to facilitate suburban development. Large tracts of land were planned, infrastructure and lots were completed, and homes were constructed that employed a limited number of model plans. The ensuing Federal Housing Administration (FHA) and Veterans Administration (VA) loan programs provided mortgages for more than 11 million new homes. These mortgages, which typically cost less per month than paying rent, were directed at new single-family construction. Intentionally or not, the FHA and VA programs discouraged renovation of the existing housing stock, multifamily homes, and mixed-use buildings.[4] These programs were incredibly effective at encouraging homeownership, and residential development exploded. Other factors that contributed to the massive growth of suburbs surrounding most major cities were the construction of the interstate highway system, the emergence of the suburban shopping center, and the early templates provided by projects such as Levittown on Long Island and Park Forest, Illinois.[5]

Once this development pattern started, builders were continually under pressure to make new homes more desirable than older homes (something akin to planned obsolescence). Until 2009, the quickest path to success proved to be a continual increase in home size, cosmetic changes, and special features such as large kitchens, master suites, and larger garages that responded to the nation's love affair with their automobiles (see Chapter 3 for more on house size). Thus conventional suburban developments over time resulted in homes of increased size that were placed farther away from urban centers, which increased Americans' dependence on automobiles. This has led to the pejorative term "sprawl."

New Urbanism Emerged in the 1990s

Land development is fraught with an often paralyzing hot-button conflict: "smart growth" versus "sprawl." The essence of the debate is that a growing number of smart-growth advocates consider typical suburban development unsustainable because it uses too much land, consumes excessive road and service infrastructure, exacerbates congestion, and delivers lower quality of life as communities are too car dependent and become socially disconnected. In contrast, advocates for conventional suburban development cite a famous line from Herbert Gans, a prominent American sociologist, "If suburban life is so undesirable, the suburbanites themselves seem blissfully unaware of it."[6] The smart-growth community offers an alternative land development strategy called New Urbanism or Traditional Neighborhood Development (TND). This development strategy promotes higher density,

---

[4] Andres Duany, Elizabeth Plater-Zyberk, and Jeff Speck, *Suburban Nation: The Rise of Sprawl and the Decline of the American Dream* (New York: North Point Press, 2000).

[5] Robert Bruegmann, *Sprawl, A Compact History* (Chicago: University of Chicago Press, 2005).

[6] Joel Kotkin, *The Next Hundred Million, America in 2050* (New York: Penguin Press, 2010).

mixed-use developments that encourage walking, greater diversity of residents, mass transit access, and more efficient use of land.

Advocates of new urbanism came from early dissenters of conventional suburban development in the 1970s and 1980s. They embraced the urban visions and theoretical models for the reconstruction of the "European" city proposed by architect Leon Krier and the "pattern language" theories of Christopher Alexander. This movement was formalized in 1993 when the Congress for New Urbanism was founded in Chicago.[7] New urbanism advocates espouse a number of fundamental rules as an alternative to sprawl:[8]

> ***Discernible Center.*** This is the focal point for a community where people can shop, meet, and engage in common activities. It may be a square or shopping center and often includes a "town hall."
>
> ***Five-Minute Walk.*** Most dwellings are located within a short walk from daily needs of life, including opportunities for work and shopping to minimize automobile trips and encourage social interaction.
>
> ***Street Network.*** The street pattern employs a grid that offers numerous paths connecting relatively short blocks to each other to relieve congestion and maximize pedestrian choices.
>
> ***Buildings Located Close to Narrow Streets.*** Outdoor spaces of greater interest are created with houses placed close to narrow streets compared to the wide streets with long setbacks typical in suburban neighborhoods.
>
> ***Mixed Use.*** A variety of commercial and residential construction enables diversity of population and human activity.
>
> ***Special Sites for Special Buildings.*** Certain prominent sites at the termination of street vistas or in the neighborhood center are reserved for civic buildings used for meeting, education, and religious or cultural activities.

An impressive and growing list of TND communities apply most or all of these rules. A partial list of communities includes:

- **Birkdale Village north of Charlotte, North Carolina**
- **Celebration, Disney's development south of Orlando, Florida**
- **Civano in Tucson, Arizona**
- **Kentlands in Gaithersburg, Maryland**
- **Mesa del Sol planned near Albuquerque, New Mexico**

---

7 "New Urbanism," Wikipedia, August 16, 2010, http://en.wikipedia.org/wikipedia.org/wiki/New_Urbanism

8 Duany et al., *Suburban Nation.*

- **Mountain House between Tracy and Livermore, California**
- **New Town at St. Charles within the city of Saint Charles, Missouri**
- **Noisette in Charleston, South Carolina**
- **Orenco Station in Hillsboro, Oregon**
- **Playa Vista in Los Angeles, California**
- **Prospect New Town in Boulder County, Denver**
- **Seaside in Walton County, Florida**
- **Stapleton in Denver, Colorado**
- **Symphony, planned to break ground north of Indianapolis, Indiana**
- **The Cotton District in Starkville, Mississippi**
- **Verrado in Buckeye, Arizona, 25 miles south of Phoenix**

Conflict between Conventional Suburbs and New Urbanism

It is useful to compare the diverging philosophies for land development from a business perspective. New urbanism advocates contend that it is much more profitable to develop traditional neighborhoods than conventional suburban projects. Developers have a greater opportunity to assess this claim by observing the market response and potential business outcomes to the growing number of TNDs across the country. Assuming developers are interested in maximizing profits, it would be logical to expect a shift to TND projects if there are clear signals of consumer preference and reduced development costs. Table 2.1 provides some perspectives on why the decision may not be so clear cut for mainstream developers.

A Word about Preaching

I have the opportunity to give presentations on energy-efficient and environmentally friendly construction practices to many different audience groups. When presenting to environmental professionals, I often start with a few questions. How many of you never drive over 55 miles per hour? How many of you never take long showers? How many of you have more than one child? Inevitably, a large number of hands go up, responding affirmatively to each of these questions. This demonstrates their routine behavior often conflicts with their staunch environmental protection beliefs based on personal preferences: driving cars on highways at 65 miles per hour, knowing a slower speed would result in significantly less vehicle emissions; taking long showers, knowing it consumes more water; or having multiple children, knowing population growth may be the biggest environmental challenge facing our planet. In fact, each of us draws a different line in the sand when it comes to our strong convictions. This suggests few, if any, of us have a compelling moral base for preaching specific

Table 2.1: New Urbanism versus Suburban Development

NEW URBANISM DEVELOPMENT PERSPECTIVE	CONVENTIONAL SUBURBAN DEVELOPMENT PERSPECTIVE
"Sprawl" is not sustainable.	Malthusian theory predicted that population growth would exceed the capacity of the agriculture infrastructure and lead to a world shortage of food. However, this theory failed to account for technical innovation and never materialized. Similarly, new developments in how people work, travel, and harvest renewable power can mitigate excessive congestion and energy use associated with suburban development.
Increased density encourages social interaction and caring for neighbors.	People often do not make great neighbors because individual interpretations of property ownership entitlements can become a slippery slope. One person's freedom to leave his or her dog barking all night outdoors, have frequent loud parties, or to clean infrequently can lead to loud noise levels and bug infestations that profoundly disrupt neighbors. The permutations of potential behaviors that can be perceived as disturbing increase with density.
Garages accessed from rear alleys encourage social interaction and more attractive street architecture.	Homeowners feel less safe with rear alley access due to additional exposure to crime.
Narrow streets with short setbacks create more interesting, well-defined outdoor spaces.	Subdivision developers' hands are tied by local code restrictions; they would prefer narrower streets to reduce infrastructure costs.
Lack of a town center within a short walk creates excessive dependence on cars.	Suburban residents vote with their pocketbooks and have a strong preference for diverse shopping accessible by car, including much lower prices than often are available in urban shopping centers.
Increased social interaction helps form communities.	Some studies suggest urban residents keep more to themselves, and it is not clear that suburban residents are not engaged to their satisfaction.
Quality of life is better in more diverse developments.	Reasons for preferring suburban communities often include better schools, less litter, fewer bug and rodent infestations, and less crime that surpercede concerns about diversity.

behaviors. More important, preaching is not required to achieve a desired outcome. Rather than dogmatically preaching "green" building practices, I find it is much more effective to demonstrate how these practices are more profitable. And they are!

With the intent of full disclosure, I personally tilt toward many of the concepts advocated by the new urbanism camp. However, as opposed to those who adamantly preach TND practices,

I am willing to compromise on many of the core principles. The "devil's advocate" perspectives presented in Table 2.1 make me more tolerant of the business decisions driving many conventional suburban developers and home buyers. The goal of this chapter is not to advocate for one development pattern over the other but to identify recommendations that can apply to both development camps. Both perspectives are important, and I choose to let the marketplace and government policy navigate this highly charged issue.

Why Sustainable Land Development Is Broken

There are important design features and practices that should be included in any sustainable development, regardless of its broader design philosophy. But they are not. Some of these key features and practices are highlighted in the following three case studies.

Case Study 1: A Drive Through the California Desert

After arriving for work at the California Energy Commission in June 1981, my first travel assignment was to represent the commission at the dedication of a new central receiver power station, called Solar One, near Barstow, California. This was a large 10-megawatt solar plant that used more than 1,800 mirrors focused at one point atop a 10-story tower where extremely hot, concentrated solar radiation would superheat water circulating in a loop from a storage tank. The superheated water would run a steam generator similar to a conventional power plant. I was very excited to make my maiden voyage as a new employee and see this innovative technology. But I digress; this story has nothing to do with this advanced new solar system.

I arrived at the airport near Barstow, got into a rental car, and proceeded to drive to the remote Solar One plant location. It was a typical California high-desert summer day with temperatures hovering near 115° F. In other words, it was crazy hot for a transplanted New Yorker. As I started my drive, I couldn't help noticing a disturbing design flaw with the relatively new housing developments in the area. As a solar architect, I was appalled that these homes were on an essentially flat landscape with no compelling views and that they completely ignored solar orientation. This was evident by the large window areas with no overhangs facing all compass directions (north, east, south, and west). The intense energy of the sunlight entering these homes was palpable just driving by them. This is incredibly inappropriate design because as much as 60 percent of space cooling loads can be attributed to incident solar heat gain. The only solution for the occupants of these homes was incredibly thick blinds (which begs the question, what was the purpose of the windows in the first place?) and brute force air conditioning. What a waste; it was equivalent to snubbing your nose at Mother Nature every hot day. Whatever good environmental benefit could be attributed to the new solar central receiver power plant I was driving to visit was being squandered many times over by these desert homes that completely disregarded the daily sun path and its impact on natural comfort.

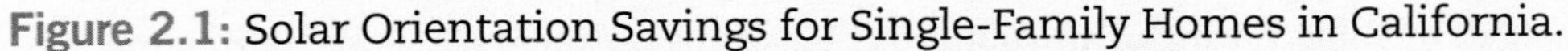

Figure 2.1: Solar Orientation Savings for Single-Family Homes in California.

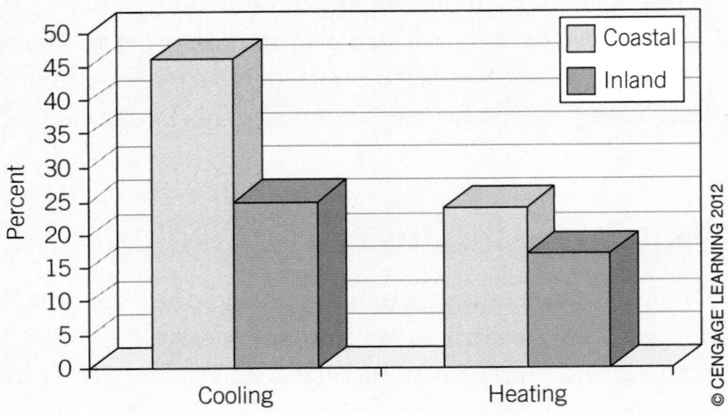

Source: "Evaluation Measurement and Verification of California ENERGY STAR Homes, Strategic Research and Evaluation," ENERGY STAR for Homes: California Utilities Evaluation Studies, August 16, 2010, http://www.energystar.gov/index.cfm?c=pt_reps_new_construction.pt_reps_partner_meeting_2007.

Imagine giving each of those Barstow homeowners the opportunity to choose how much *free* space heating and cooling energy they wanted to throw away. I say "free" because it costs virtually nothing to integrate solar orientation into a subdivision plan and into house design. I suspect that most, if not all of the owners, given this choice, would have elected to not give away any free natural cooling. But, in fact, the housing industry across the country routinely ignores development patterns that could reduce space heating and cooling costs approximately 25 percent at no extra cost. As an example, Figure 2.1 shows the heating and cooling savings for both coastal and inland locations in California by taking advantage of a southern orientation.[9]

Case Study 2: Conventional Suburb Developer in California

My wife and I moved to Sacramento, California, from Long Island, New York, in June 1981. Getting on the plane to California was one of those unique times as an adult when we didn't possess a single key because we did not own a car or have a place to live. As soon as we settled into a hotel in Sacramento, the process of shopping for apartments took top priority. All of our possessions were on route with a mover, and we had to have a place secured

[9] "Evaluation Measurement and Verification of California ENERGY STAR Homes, Strategic Research and Evaluation," ENERGY STAR for Homes: California Utilities Evaluation Studies, August 16, 2010, http://www.energystar.gov/index.cfm?c=pt_reps_new_construction.pt_reps_partner_meeting_2007

within a week for their arrival. Thus, we had a bit of time pressure just to make the search interesting. It turned out that every day that first week was over 100° F. Welcome to a typical central California valley heat storm. A few experiences getting into our car after it sat in the sun and reached 150° F (personal estimate) firmly established the value of shade trees in our minds.

In a seemingly endless search across the Sacramento area, we struggled to find the right place. Ultimately, we found an apartment in a development called Woodside. It was part of a condominium project by Sacramento developer Robert C. Powell. This place was different. All the units were nestled among large mature trees, including preserved live oak and walnut trees along with an impressive array of other species that had been effectively planted early during the construction process. With all of the natural shading, the microclimate felt about 20 degrees cooler (a dry 85° F is much cooler than a dry 105° F). In addition, the project was enhanced with beautiful walking trails, lush landscaping, a beautiful stream system running through the property, and a full-featured community center with meeting rooms, several pools, and tennis courts. This was more beautiful than most vacation resorts my wife and I had visited. The price was right, and we signed the rental agreement.

Subsequently, we came to appreciate that Robert C. Powell applied the following basic formula to all of his developments in Sacramento:

> ***Invest in a Proven Landscape Architect.*** Professional landscape design makes a profound difference by ensuring a diverse pattern of ground cover, shrubs, turf, and trees appropriate to a specific climate and spacing requirements that accommodate mature growth.
>
> ***Invest in Implementation of the Landscape Plan.*** Once you have a good landscape design, it will typically entail added costs to preserve large trees, add an extensive amount of newly planted trees, and provide the right combination of turf, ground cover and shrubs throughout the development along with good quality hardscaping (e.g., trails, accent paving, attractive lighting). But it completely transforms a community.
>
> ***Ensure Persistence of Landscaping and Home Maintenance.*** Powell developments typically allocated maintenance of all front yards and open space to the Home Owner Association (HOA). This ensured all of the front landscaping was professionally nurtured to a mature lush appearance throughout the neighborhood. In addition, HOA dues ensured roofs were replaced and siding painted on schedule. This helped enhance property values with neighborhoods that kept looking better with age and cost-effectively eliminated annoying burdens of homeownership.

All of Robert C. Powell's existing developments effectively served as showrooms depicting unique lush aesthetics and well-maintained homes. Even if a site didn't have large trees, fast-growing trees were planted throughout the neighborhood as early as possible in the site development process. Thus, even when one of the largest developments in the region, Gold River, on the less-preferred south side of the American River, opened during a major real estate slump, long lines formed to buy homes. Gold River has gone on to be another impressively successful development. The trees have matured, the landscaping is lush throughout, generous walking trails are widely used, and there is extensive open space along the beautiful American River parkway. Although this and other Powell developments are far from perfect, they demonstrate the critical role effective landscaping, solutions for ensuring proper maintenance, and community space play in successful development. All of these practices are underutilized in the development process.

Case Study 3: A Profitable Development No One Copied

Michael and Judy Corbett were not your typical land developers. They had no experience developing land, but they had the persistence and vision to execute one of the most impressive model solar communities in the nation. Over the strenuous objections of city planning and public works departments and the Federal Housing Administration, the Corbetts accomplished a profoundly important design that has paid off in unexpected bonuses for the environment and for the residents of Village Homes.[10] Breaking ground in 1975, the Corbetts started the process of converting a 68-acre tomato farm on the western end of Davis, California, about 15 miles west of the state capital, into a community with 244 housing units: 222 single-family homes and 22 apartment units. In addition, there is a small 4,000-square foot commercial center, 12 acres of common agricultural land, two village greens, a swimming pool, community center building, one restaurant, a dance studio, and a day care center (Figure 2.2).

Much of the fighting with the local planning officials revolved around the narrow streets and a natural drainage system. Narrow streets were designed to discourage traffic, allow trees to more effectively shade the road, and provide a more natural look and feel to the community. City planning officials' concerns about the ability for emergency vehicles, particularly fire trucks, to negotiate the streets were met by mandatory 30-foot easements on both sides of the street. A network of drainage swales was a natural alternative to expensive and wasteful storm drains.[11] These swales collect water at the rear or side of properties and function like seasonal streambeds with rocks, bushes, and trees. This results in properties sloping away from rather than toward the streets.

Open space was a critical element to the success of Village Homes, accounting for 25 percent of the site. This includes the streets, agricultural land, village greens, orchards, common areas, playgrounds, community gardens, bicycle and pedestrian paths, and land around the drainage swales. All of this space has effectively facilitated substantial social connections within the community that many report as an important attraction to living there. Based on one study,

[10] Bill Browning and Kim Hamilton, "Village Homes, A Model Solar Community Proves Its Worth," *The Context Institute, Designing a Sustainable Future* 35 (Spring 1993).

[11] Ibid.

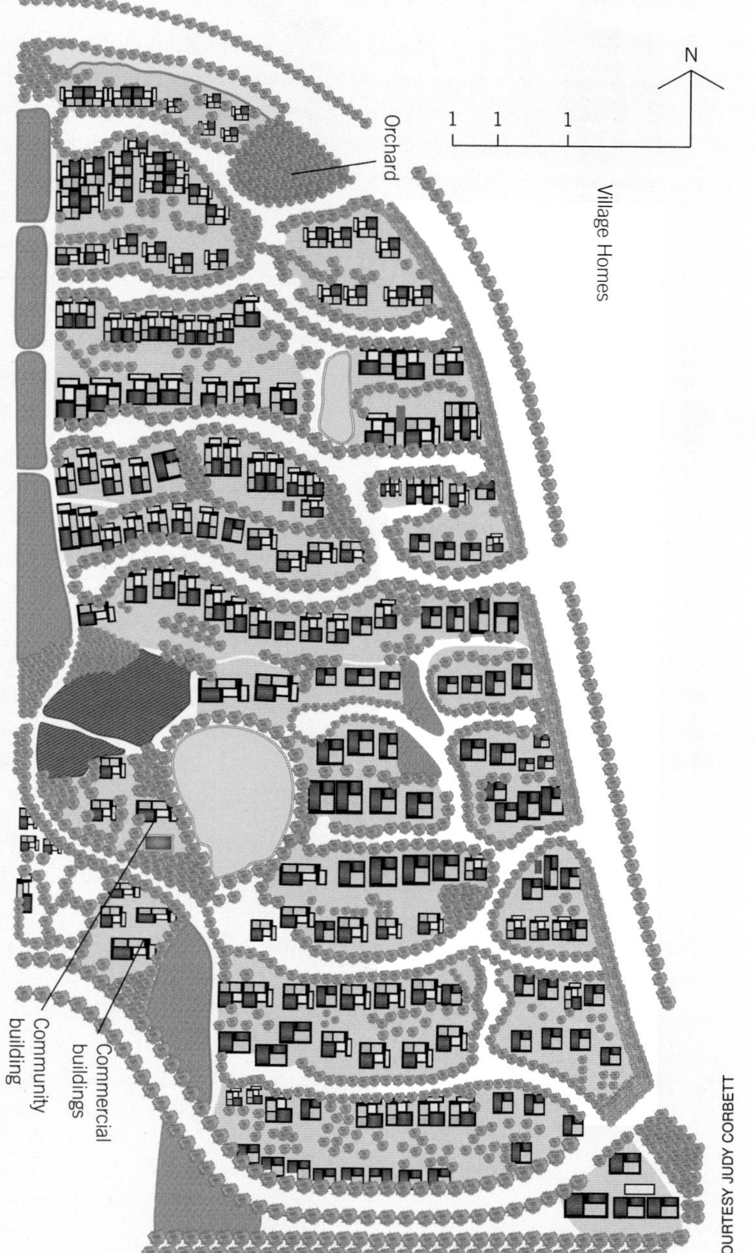

Figure 2.2: Village Homes Site Plan.

"most residents appreciated the unique social life of the neighborhood, including its communal open spaces, appropriateness for children, and opportunity for social contacts . . . people living in Village Homes had twice as many social contacts as people living in other parts of Davis."[12]

One of the most significant design elements of Village Homes was that the Corbetts demonstrated that east-west-oriented streets could easily be integrated into a development plan without any additional cost or aesthetic compromises. The result is that virtually every lot features a north-south-oriented home. This orientation was fully optimized by requiring a passive solar design for each house. "Solar energy contributes between 50 percent and 75 percent of the heating needs. All of the houses have 60 percent or more of their glazing on the south side. The most basic solar features are calculated overhangs on the south facades, which shade the houses in the summer but allow sun into the homes in the winter. They also have extra insulation in roofs, and concrete slab construction for thermal mass. Almost all of the homes have solar hot water systems with collector panels on the roof."[13] Thus all homes have natural comfort design (see Chapter 3). I have chosen to use "natural comfort" as a much less contentious term for "passive solar." This is due to the negative perception of some highly eclectic early designs.

Clearly this is a unique approach to developing land. But is it profitable? One study from 1991 indicated that houses in Village Homes were selling for a premium of $11 per square foot over other Davis developments.[14] Another study suggests that Village Homes had become the most desirable subdivision in Davis, with homes selling at a $10 to $25 per square foot premium in 30 percent less market time.[15] These studies suggest a strong market preference for this unique approach to development. Unless you visit Village Homes personally, it is difficult to appreciate the uniquely superior performance of this neighborhood. Beyond its natural beauty, there is an intangible sense of neighborhood. Residents are interacting in gardens, orchards, playgrounds, and open green spaces. You feel the incredible difference natural shade provides for cooler streets. Walking inside homes with proper orientation, you realize how comfortable spaces can be without artificial air conditioning even during hot summer days in the Sacramento Valley.

Village Homes is not a TND community, and it goes against many of the popular principles advocated by new urbanism and the smart-growth planning movement.[16] It features long curving streets with cul-de-sacs rather than a grid of short streets. It does not have a major town center, and as a result, residents cannot access the needs of daily life within a 5-minute walk. There is no significant diversity of income groups. Density is fairly similar to most conventional suburban developments. But by most measures and resident feedback, Village Homes is an incredibly successful development. Various detailed studies have offered suggestions for improvement, but I submit that there is no such thing as the perfect development. This is truly an important success story in land development.

---

[12] Mark Francis, "Village Homes: A Case Study in Community Design," *Landscape Journal* 21 (January 2002).

[13] Browning and Hamilton, "Village Homes."

[14] Ibid.

[15] Francis, "Village Homes: A Case Study."

[16] Ibid.

If significant evidence and empirical observations demonstrate the increased value and livability of Village Homes, why is this development and its successful design philosophy relegated to an interesting aberration? The only answer I can surmise is that it is easier, more convenient, and adequately profitable to follow the conventional suburban design template. Moreover, just one development may not provide enough "proof of concept" for other developers. Maybe Village Homes only appealed to a niche customer and this success would not transfer to other larger projects. I could go on, but you get the idea. Mainstream developers appear to consider the success of this historic an exception, and continue with business-as-usual development practices. In the process, they have ignored the natural comfort, open space, landscape planning, and community features that make Village Homes one of the most livable communities in the country. Now it is time to take another look.

How to Fix Sustainable Land Development

As indicated earlier, the debate between new urbanism, conventional suburban development, and all the variations in between will remain beyond the scope of this book. Many will find this a critical shortcoming to this discussion, but I understand the valid points on all sides and choose not to preach on this subject. Moreover, with the increasing number of TNDs around the country, their performance can be fully evaluated and market forces should naturally influence developers where a compelling business case can be effectively demonstrated. Alternatively, policymakers may choose to step in and mandate specific development patterns associated with new urbanism principles. The following retooling recommendations are removed from this debate and apply to developers regardless of their preferred design philosophy. They are critically important because they represent the opportunity to get land use patterns right before they are locked in, often for centuries.

Optimize the Street Layout for South Orientation

With the advent of modern heating and air conditioning systems, commonsense practices for optimizing solar heat gain in winter and minimizing solar gain in summer have long been forgotten. Brute force has won. Chapter 3 details natural comfort design practices for new homes, but it all begins with the site planning process. In many markets, proper solar orientation can save about 25 percent of home cooling and heating loads. Now is the time to assume home buyers do not want to waste this free energy, particularly with new sensitivities to rising energy costs, energy independence, and environmental protection. The land development planning phase provides the critical opportunity to get this right. Thereafter, it is lost forever, or at least until the development is leveled. If there are no compelling views or topographical considerations, south orientation should be maximized with streets oriented predominantly east-west. Where the front of the home faces south, the home designs can employ rear garages to maximize south-facing glass. Where the rear of the home faces south, home designs can use front garages, but more attractive front elevations and social connections will benefit from detached rear garages front porches.

Invest in Open Space

It is difficult to give up revenue-generating land for open space, but it is critical for sustainable development. Lot sizes can often be shrunk to accumulate meaningful open space, but developers should be willing to lose some lots if necessary. Desirable open space functions include park areas, bicycle and walking trails, playgrounds, common agricultural land, community gardens, sport courts, central greens, and a community center. The resulting increase in value of all lots should provide an impressive return on investment and make up for lost lots. A great example is the historic city of Savannah, Georgia. Open park squares were included every few blocks and significantly enhance social activity and neighborhood beauty compared to a typical grid of blocks without open space. The resulting increase in property value for the entire city, and particularly lots bordering the squares, should substantially exceed the value of the lost development space.

Invest in Trees and Landscaping and Experts to Do It Right

With apologies to all my architectural colleagues, the secret of residential developments that stand the test of time is that mature trees and landscape design often trump house design. However, emphasis appears to be placed on factors driving the initial sale such as house architecture and highly decorated models rather than on long-term customer satisfaction of living in expertly landscaped neighborhoods that get better with age. This entails a painful change from conventional clear cutting to preserving healthy and desirable mature trees and shrubs and then integrating them into a lush landscape plan for the community. Skilled landscape architects have a profound impact on a development's quality and ambiance. They are worth the investment.

Invest in Quality Hardscaping

Hardscaping involves all exterior architectural features used in public spaces. This includes everything from street signs, street lights, water features, fire hydrants, sitting benches, fencing, walkway materials and accents, to entries, porticos, and trellises. Again, an expert landscape architect should be utilized to select high-quality materials and products, develop an integrated design, and suggest strategic accents. The positive impact on the perceived value and quality of the development will be substantial.

Set Up Home Owner Association Maintenance Covenants

Homeownership is hard. Many buyers are ill prepared for the discipline necessary to maintain their property and homes. To lessen this burden, I recommend adopting Powell's practice of setting up mandatory annual fees to a Home Owner Association (HOA) for maintenance of all the front yards, common spaces, and exteriors. This ensures yards, open spaces, and homes are continually maintained throughout the neighborhood. Moreover, it creates the effect that homes are sited in one continuous lush landscaped setting where the shrubs and trees are expertly selected for durability and appearance appropriate to the local climate. The positive impact on resale value and relief from maintenance will increase customer satisfaction and enhance the developer's reputation. This is a win-win, low-cost action for developers.

Lock In Integrated Quality Design Features with a Plan Book

Disjointed architectural treatments can compromise a neighborhood's appearance and value. Old European villages that utilize consistent classic architecture with rich details have timeless beauty. Sustainable developments should strive for a similar goal by using a plan book to effectively establish a preferred architectural style and set of details for each community. This includes exterior architectural elements such as fencing, porches, exterior columns, pavement, decking, trellises, and exterior lighting. The plan book should also be used to ensure layouts that avoid "pig snout" front elevations in which two- and three-car garages dominate the street experience. Yes, American homeowners must have their garages, but there are alternative design solutions. For instance, rear garages accessed by side driveways or rear alleys completely free up the front elevation for each lot. Front porches should be a mandatory feature to encourage social interaction and provide weather protection at the entrance. The entire community will benefit from increased property values associated with consistent quality design features.

Establish Special Reasons to Live in a Development

In November 1994, I moved from California to the east coast for a new job in Washington D.C. The home search encompassed the massive metro D.C. region. With two young children and personal concerns about quality of schools, drinking water, safety, and competent government services, the district itself was out of consideration for our family. Our search focused on the suburbs, but we refused to live in a typical tract neighborhood with no special features. Ultimately, we found a beautiful neighborhood with large mature trees, natural drainage, beautiful open spaces including a pond, forests, and a small park, nearby large regional parks, neighborhood swim and tennis center, great schools, good shopping and services within walking distance, and easy access to commuter train service. I suspect my selection process is more critical than that of the average home buyer, but new developments must include several special features that help make the development a compelling place to live. Here is a partial list of features that qualify as special:

- **Special vistas**
- **Unique location (e.g., proximity to work centers, parks, mass transit, water, shopping)**
- **Recreational amenities**

- **Desired open space (e.g., parks, community agricultural land, trails, preserved forests)**
- **Town center**
- **Nearby medical services**
- **Covenants that protect land value (e.g., maintenance requirements, architectural restrictions)**
- **Featured architectural style**
- **Advanced technology infrastructure**
- **Green development and construction practices**

If you are going to develop a piece of land, make it special!

Chapter 2 Review

So What's the Story?

Once land is formally developed, its design and function are virtually locked in for centuries, for better or worse. Thus there is substantial pressure to get it right with projects that are aesthetically pleasing and financially viable. The sustainable land development story can be summarized as follows.

- **What It Is.** Sustainable land development is a process for securing, designing, and preparing land for home construction that is aesthetically, functionally, and socially prepared to last for hundreds of years.
- **How It Got Here.** Government policies and national mortgage programs led to a transition from predominantly urban to predominantly suburban development following World War II.
- **Why It's Broken.** Residential developments do not consistently optimize opportunities for social connections, expertly designed green space, and natural comfort homes.
- **How To Fix It.** At every step of the land development process, lock in open space, natural comfort, lush landscaping, and quality hardscaping.

The details are included in Table 2.2.

Table 2.2: Sustainable Land Development Summary

What It Is	How It Got Here	Why It's Broken	How To Fix It
Five-step process that captures opportunities to integrate open space, natural comfort, landscaping, and hardscaping: 1. Research 2. Land acquisition 3. Architectural plans 4. Entitle land 5. Site work	Urban development dominated to the end of WWII. Suburban development dominated after WWII. New urbanism emerged in the 1990s. Conflict has grown between conventional suburban development and new urbanism.	Lack of effort to integrate solar orientation in subdivision planning Lack of investment in quality landscaping and ongoing resistance to maintaining and nurturing the natural environment Lack of open space and features that encourage community interaction Lack of consistent architectural elements and style	Optimize the street layout for south orientation if there are no compelling views or topographic constraints. Invest in open space. Invest in trees and landscaping and experts to do it right. Invest in quality hardscaping. Set up HOA maintenance covenants to nurture and maintain open spaces and landscaping. Lock in integrated quality design features with a plan book. Establish special reasons to live in a development.

3

Good Housing Design: Design Trumps Everything

Good Housing Design: Process, Goals, and How Goals Are Achieved

Process. *Good housing design is a three-step process that seeks to optimize aesthetics, function, and systems for a specific location and budget.*

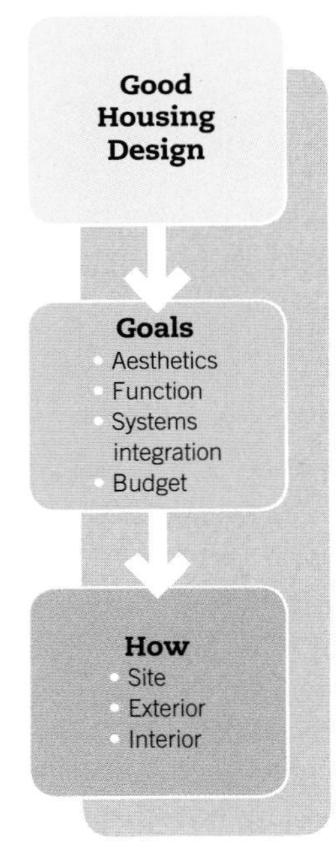

Step 1. Programming. *Design requirements are clearly defined. In the case of custom homes, these requirements are based on client interviews. For "spec" homes, builders, architects, and designers have to rely on market research and intuitive judgments. The resulting design parameters commonly include number and types of rooms, number of floors, spatial requirements, amenities, furnishing needs, architectural features, and in some cases special performance requirements such as green features and right-sized homes.*

Step 2. Design Schematics. *Design requirements are converted into sketch plans, elevations, and in many cases three-dimensional drawings, often with the aid of CAD (computer aided drafting) software programs.*

Step 3. Design Development. *The original design schematics are benchmarked to the programming*

© iStockphoto.com/alexsl. © iStockphoto.com/smanter.

requirements and budget and refined through a series of revisions. This continues until a "best solution" final design is approved by the client for custom homes, the builder for spec homes, or the designer for stock plans. Architectural models may be prepared to ensure there are no surprises and to facilitate client decisions, but three-dimensional CAD drawings have diminished the need for this costly effort.

With approval of a final design, the process is complete and ready to move on to preparation of construction documents (see Chapter 5). At this point the design is substantially set.

Goals. *The first obvious goal for good housing design is that aesthetic requirements must be satisfied. Buying a home is an emotionally driven process, and the right look and layout are critical to success in the housing industry. Beyond look and feel, good designs provide great function in terms of convenient flow, adequate room sizes, generous storage, good natural and artificial light, privacy, weather protected entries and outdoor spaces, indoor-outdoor linkages, durability, and ease of maintenance. Houses incorporate diverse systems that must all be integrated, including structural, comfort, electrical, plumbing, entertainment, water management, and disaster resistance. Finally, good design has to address all of these concerns within an established budget.*

How. *Good housing design goals are met from the outside in. This begins by addressing a full array of site concerns, including topography, drainage, solar orientation, special views, preserving desirable trees, and landscaping. Exterior architecture should provide integrated four-facade interest, minimum maintenance, effective drainage of rain and snow from the home, maximum daylight, consideration for solar orientation including passive solar design and shading, adequate weather protection at entries and exterior spaces, and optimal use of indigenous materials. Interior design should provide optimal function within minimal space, interesting and convenient flow, generous storage, room for likely furnishings, appropriate and durable materials, effective lighting, and appropriate quality trim and finishes. Housing* *designs that satisfy these goals will be substantially matched to site and climate considerations, can be constructed with less waste, provide greater function with less square footage, comprehensively address all critical building systems, and minimize homeowner durability concerns. Delivering this type of home on a consistent basis is good for business and for homeowners.*

What Is Good Housing Design?

The title of this section poses one of those million-dollar questions. Good design, often considered more an art than a science, is open to many interpretations. However, I am undeterred and boldly submit five specific criteria for good design in this section. My recommendations reflect a strong belief that sustainable design can no longer be ignored because it offers superior benefits for both homeowners and builders.

A Word about Design and the Creative Process. I often wonder why many of my favorite creative artists only have short windows of brilliance directing a few great movies, releasing a few consistently strong CDs, or writing a few powerful books. I have come to conclude that the creative process is insanely pressurized and extremely difficult to sustain because it relies so much on inspiration that is extraordinarily difficult to summon on demand. Thus successful artists often experience only a limited period of success during which they create their best work, and their lifetime reputations are often based on these short windows of creative inspiration. Artists who have prolonged periods of success, I would submit, can attribute that success to a uniquely repeatable creative process: a formula or template for directing successful movies, composing hit music, or writing best sellers. Architectural design is often mistakenly grouped with these other creative processes, but in fact is a professional skill that each client expects consistent results. Patients would find it unacceptable for surgeons to have good and bad days performing medical procedures, and the same applies to clients seeking professional architectural design services. The discussion in this section seeks to identify the critical criteria that consistently yield good housing design.

Good Housing Design Criteria 1: Fit Homes to Site

Production builders typically rely on a limited number of house plans and elevations. However, even stock plans can be adapted to individual sites and include consideration of the following factors:

> ***Fit with Prevailing Views.*** Many neighborhoods have vistas (e.g., mountains, lakes, oceans, sunsets, or clusters of mature trees) or attractive amenities (e.g., ponds, trails, special landscaping, or neighborhood parks). Simply "plopping" a home on a site (technical term for house placement that ignores site factors) without consideration for these views is a lost opportunity to permanently make that home better.

Figure 3.1: House with Grade That Slopes Toward Home.

Fit with Drainage Patterns. Homes should be sited so water effectively drains away from every elevation. No exceptions! Virtually every homeowner would prefer to significantly minimize the risk of moisture damage and mold. Where natural terrain does not allow for a home to be sited at a high point, drainage swales should be designed into the grading plan to ensure positive water flow away from the home. Figure 3.1 shows a home in which the yard slopes into the home, but the land could have been contoured to create a swale that carries water away from the home.

Fit with Livable Communities. Land should be developed to encourage social interaction with pedestrian friendly streets, less dominance by automobiles, lush landscaping, walking trails, and open spaces (see Chapter 2). Home designs should contribute to this sense of community by including front porches and, where possible, rear garages.

Fit with Region. Older housing often reflects a time when high-quality materials and craftsmanship were much less expensive and designs substantially addressed prevailing weather conditions and locally available materials. Although it may be unreasonable to expect mainstream new homes to feature the same quality of materials and workmanship as older homes, it is still important to employ regionally responsive designs that address prevailing climate factors and use indigenous materials. The resulting homes will be naturally more comfortable, more durable, and look better because they are regionally appropriate.

Figure 3.2: Sun Path at Different Times of the Year.

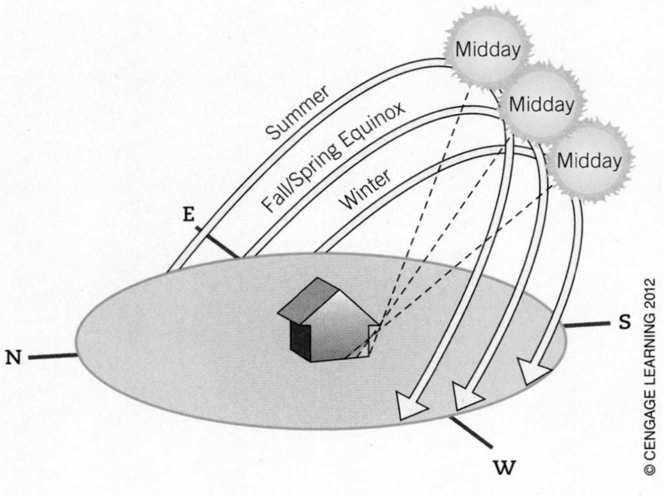

GOOD HOUSING DESIGN CRITERIA 2: INTEGRATE NATURAL COMFORT

Sustainable land development (see Chapter 2) maximizes the number of lots with a north-south orientation. Natural comfort design solutions complete the process by ensuring homes take advantage of this advantageous orientation that allows maximum control of the varying sun path throughout the year. Specifically, the sun is higher in the sky during the summer and lower in the sky in the winter. The sun path also has a much wider arc in summer than winter. As a result, the sun rises in the northeast and sets in the northwest during the summer, but rises in the southeast and sets in the southwest during the winter (Figure 3.2).

Climate-responsive designs employ properly designed roof overhangs based on the varying path of the sun. Since the sun is high in the sky in the summer when solar heat gain is not desired, it can easily be shaded with an appropriately calculated roof overhang based on latitude (commonly 2 to 3 ft in the continental United States). During the winter when solar heat is desired and the sun is much lower in the sky, sunlight can pass under that same overhang and enters the window for free winter heating (Figure 3.3).

Sunlight is transmitted in the form of radiation that is short-wave, high-temperature energy that easily passes through the molecular structure of glass. When interior materials absorb this short-wave radiation, it is emitted as long-wave heat that can no longer easily pass through the glass. Thus most of the incoming solar energy is trapped within the space until it is lost by conduction and convection through the insulated envelope, windows, and doors. This is referred to

Figure 3.3: Winter and Summer Sun Angles with Overhangs.

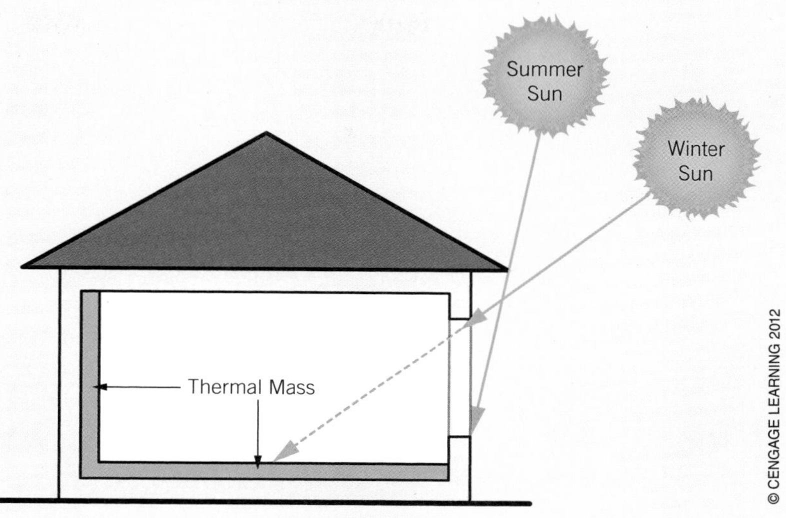

as the "greenhouse effect." Passive solar designs collect this heat in thermal mass materials that can effectively store and rerelease it over time rather than letting spaces accumulate heat to uncomfortable levels. This allows homes to get the most use of the collected solar energy because even in winter homes can overheat if the incident solar heat is not stored. In effect, thermal mass functions like a thermal battery that absorbs solar heat during the day and releases it later after the sun has set. It can be effectively integrated into the architectural design with common materials such as colored concrete or tile floors, masonry walls, and stone counters.

Different materials have different heat storage capabilities, so detailed calculations are needed to precisely size the amount of thermal mass needed for thermal control. The heat storage capacity is a function of a material's specific heat (amount of heat required to produce a unit change in temperature per unit of mass that is constant for each material), density, and mass per unit volume. When selecting a thermal storage material, the objective is usually to contain as much heat as possible per unit volume while still meeting aesthetic and cost criteria. Table 3.1 shows the comparative heat storage capacity for a variety of common construction materials. As can be seen, there is wide variation in effective thermal mass performance. However, for those that like rules of thumb, at least 5 sq ft of the most common thermal mass materials, at least 2 in. thick should be located in rooms directly exposed to solar heat gain for each square foot of south-facing glass in those rooms.

Natural cooling predominantly relies on the same south orientation and shading used for passive solar heating to block the summer sun. In addition, minimal east and west windows are critical because the low morning and evening sun is nearly impossible to control with

Table 3.1: Heat Storage Capacity of Common Construction Materials

Material	Specific Heat (Btu/lb-°F)	Density (lbs/cu ft-°F)	Heat Storage Capacity (Btu/ft³/°F)
Water	1.0	62.4	62.4
Steel	0.12	489	59
Wood, oak	0.57	47	26.8
Brick	0.20	123	25
Concrete	0.156	144	22
Expanded polyurethane	0.38	1.5	0.57
Air	0.24	489	0.018

Source: "Passive Solar Heating & Cooling Manual, Part 1 of 4," Arizona Solar Center, December 27, 2010, http://www.azsolarcenter.org/tech-science/solar-architecture/passive-solar-design-manual-intro.html

traditional overhangs. In very hot climates, west-facing windows are particularly problematic because the low sun is coincident with the hottest part of the day and can result in intense solar heat gain. As a result, comfort in west-facing rooms is often difficult to maintain even with modern air conditioning systems because the incident solar Btu load through windows exceeds the ability of the air conditioning system to remove that Btu load. Note that wing walls complement overhangs at south-facing windows by blocking the low east and west sun hitting those windows due to the wide arc of the summer sun path.

Figure 3.4 shows an infrared image that compares temperatures on sections of a concrete floor that are shaded, exposed to sunlight through a bronze-tinted window, and exposed to sunlight through a low-e window (see Chapter 4). The shaded floor is more than 15° F cooler than the floor exposed to a tinted window and nearly 5° F cooler than the floor exposed to a low-e window. This demonstrates that the most effective cooling strategy is to block the sun from entering the home, but also shows that low-e windows are effective at reducing a substantial portion of solar heat gain. However, low-e windows also block desired solar heat gain during the winter.

There are additional strategies for natural cooling beyond shading. Homes can be ventilated with cooler night air to reduce the temperature of the thermal mass (e.g., masonry surfaces and drywall). This ventilation can be achieved with mechanical fans or passively with a concept called the solar chimney. This relies on the stack effect in which the warm air naturally rises and a design that includes high operable windows (e.g., skylight, or high ceiling with pop-up roof monitors). With lower windows open during cool evenings, hot air naturally circulates up and out the high windows, creating a natural circulation flow. This is primarily

Figure 3.4: Surface Temperature of Concrete Floor with Different Windows and Shading.

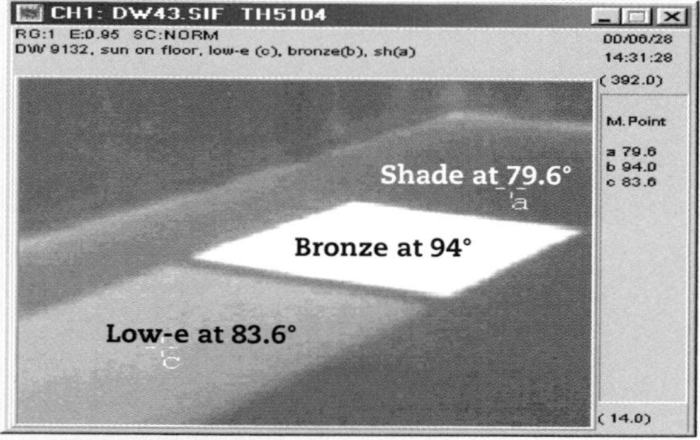

COURTESY OF ARIZONA STATE ENERGY OFFICE

effective in dry climates with significant daytime to evening temperature swings such as occur in the west and southwest regions of the United States. In these climates, it is not unusual to have evening temperatures drop to 60° F or below on days that peak at over 100° F. Natural cooling also can be provided by orienting homes to take advantage of prevailing summer breezes so open windows create a cross-ventilation flow of air. This is no longer considered acceptable in even mildly humid climates where air conditioning is considered essential for dehumidification.

As seen from this discussion, natural comfort design requires knowledge of peak sun angles for winter and summer. These can be determined with special solar tracking charts for various latitudes often found in old passive solar design handbooks. However, I recommend using the following Web sites:

- **Sun angle and weather data at http://www.Gaisma.com**
- ***Climate Consultant,* at www.energy-design-tools.aud.ucla.edu/**
- ***"Sun or Moon Altitude/Azimuth Table: U.S. Cities and Towns,"* the Naval Oceanography Portal Web site, at http://www.usno.navy.mil/usno/astronomical-applications/data-services/alt-az-us**

Finally, it should be recognized that true north can vary significantly from magnetic north and needs to be considered when specifying the orientation. Figure 3.5 shows these variations across the continental United States.

Figure 3.5: True North and North Variations in Continental United States.

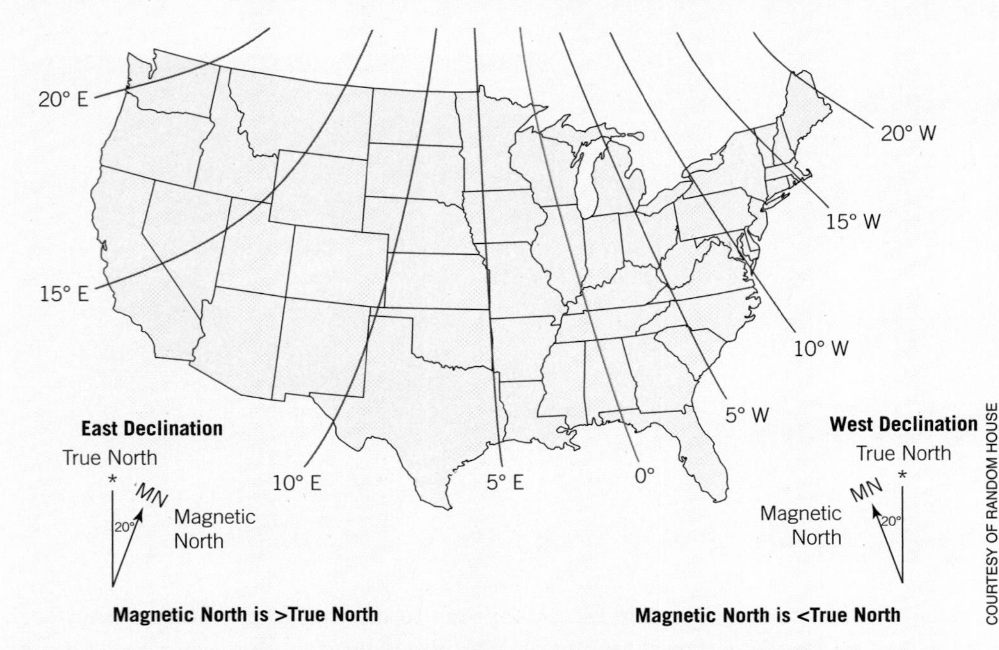

Source: Rick Curtis, *The Backpacker's Field Manual* (New York: Random House, 1998), 146.

Good Housing Design Criteria 3: Right-Sized Homes

Right-sized homes are significantly smaller than typical homes built today while retaining full function and improving performance, durability, safety, and quality. This is similar to trends in new cars, which are dramatically smaller than their bloated counterparts from the 1960s and early 1970s but deliver full spatial comfort along with substantially improved performance, durability, safety, and quality. By virtue of their smaller size, right-sized homes can help the housing industry address increasing pressures to control construction costs during the current soft market. I have consistently designed and promoted right-sized homes throughout my career as an architect and energy-efficiency program manager. I have been encouraged to see this design philosophy emerge as a movement with the highly successful *Not So Big House* series by Sarah Susanka that began in 1998.[1] For purposes of this discussion, I offer the following 10 principles for right-sized homes.

[1] Sarah Susanka, *The Not So Big House Book* (Newtown, CT: Taunton Press, 1998).

1. ***Open Layouts.*** Compartmentalized floor plans that break up large spaces into small rooms make homes feel smaller. Visually and functionally linking spaces creates longer fields of view, shared common spaces with greater functionality, and an overall sense of size much greater than the individual areas within the open space. For some reason, the western region of the country has been much quicker to embrace this design concept in mainstream housing. However, there are signs that open layout designs may be spreading across the country.

2. ***Eliminate or Combine Rooms.*** A related concept to open layouts is to eliminate rooms that are little used or to combine them with other similar rooms into larger spaces for increased function and perception of size. Common examples include eliminating formal dining rooms and living rooms and opting to just provide a larger eating area combined with a "great room" off the kitchen. Great rooms also often include other spaces such as entries, office nooks, and special-use alcoves.

3. ***Minimal but Generous Circulation.*** It is probably safe to assume that home buyers are not looking for more hallway space in their homes. Reducing circulation space is an obvious right-sized home recommendation with no downside. What is a seemingly contradictory second part to this recommendation is to make the remaining circulation generous. Nothing conveys "smallness" more than a narrow hallway; a long narrow hallway is deadly. It is critical that a hallway have either more than minimum width or strategic bump-outs at room entries or linen closets to create a sense of space. Another way to make a hallway appear larger is to add daylight (a technical term for adding a window).

4. ***Varied Ceiling Heights.*** When ceiling heights are all one level, the lack of visual interest makes spaces appear smaller. As will be discussed later, increased ceiling heights have become more common. This trend can be used to strategically vary ceiling heights at room transitions to make them feel bigger. Although increased house volume adds to the heating and cooling load, it should not have a significant impact.

5. ***Built-In Furniture.*** Luxury boats provide great examples of how small spaces can live much bigger through the use of built-in furniture. Similarly, I have consistently used extensive built-in furniture throughout my house designs, both new and retrofit. I have personally estimated that this strategy consistently made my projects live about 25 to 30 percent larger than their actual square footage. Significant opportunities for extensive use of built-in cabinets include cabinet/shelf units, hutches, corner cabinets, hall linen cabinets, entertainment centers, dresser/shelf units, closet storage systems, desk/shelf units, large bathroom vanity cabinets, recessed ironing boards, laundry cabinets, window seats, mud room bench and storage units, and garage storage units. Built-in cabinets have such a great impact because they eliminate significant amounts of wasted space on either side of, and behind, free-standing furniture and often extend from floor to ceiling. As a result, they provide much greater storage and can eliminate clutter. Homes live substantially bigger without clutter. That's just a fact of life.

6. ***High-Quality Trim, Hardware, and Finishes.*** Size should never be increased at the expense of quality fit, trim, and components. Important finish details include baseboards, window and door casings, chair moldings, door hardware, cabinet hardware, countertops, plumbing fixtures, bathroom accessories, and closet storage systems. Rich interior appointments more than compensate for reduced size and make a space feel luxurious.

7. ***Effective Use of Colors.*** I usually go to Restoration Hardware to buy paint when my wife and I remodel our homes (which we always seem to do). Paint there is at least twice the price per gallon of paint at a big box hardware store, so why do I go there? As an architect with lots of design experience, I know the importance of color. Just a slight shade difference or ineffective contrast can have a profound impact on the overall house design. I want zero risk of dissatisfaction, and Restoration Hardware has room after room in their stores painted with a premium color palette developed by highly expert interior designers. The ability to see entire rooms with specific color options exactly like they will appear in my home provides me with critical peace of mind regarding my final choices. It is well worth twice or even three times the price. After all, the cost of the labor far exceeds the cost of the material. Moreover, the right colors are critical to effectively highlight higher-grade trim and details. I am getting extra value for the investment in quality details while making my home feel richer and live bigger.

8. ***Daylighting.*** Dark spaces feel smaller. Good artificial lighting is essential (see next recommendation), but natural daylight is better. People prefer daylight because it feels more comfortable and healthier. More important, for right-sized homes, daylighting makes spaces feel bigger. Effective daylighting is enhanced with natural comfort design in which a north-south orientation provides the maximum number of windows that can deliver controlled sunlight: south windows with effective overhangs, north windows that deliver mostly diffuse light, and minimal east and west windows with difficult to control low morning and evening sun angles.

9. ***Effective Lighting Design.*** I recently took my first flight on Virgin America from Washington D.C. Dulles Airport to San Francisco. I usually am just focused on getting to my seat and getting to work when I board a plane, but I was totally distracted from my normal business mind-set the moment I stepped on the Virgin America plane. The entire experience was different in a good way. The same basic plane architecture was impressively enhanced with an advanced lighting design featuring color and softer ambiance. Virgin America understood that good lighting design can create a better experience and took the opportunity to enhance what is normally a drab mood aboard aircraft.

   Good lighting design also is an important technique that can transform homes so they live bigger and feel richer. I capture this opportunity on my own architectural projects by devoting significant design effort and at least one plan for each level just to the lighting design. Figure 3.6 shows how a simple combination of direct and indirect lighting in a

Figure 3.6: Advanced Lighting Design Example.

Source: "Saratoga Energy Efficient Home," Lighting Research Center, Rensselaer, December 22, 2010, http://www.lrc.rpi.edu/programs/lightingTransformation/residentialLIghting/saratoga-Home/index.asp

family room can significantly enhance the ambiance of the space. As a note of interest, all of the lighting in this figure is also high-efficacy using about 75 percent less wattage than with typical incandescent lighting.

10. ***Indoor-Outdoor Linkages.*** Indoor living spaces visually linked to outdoor living spaces feel bigger by visual extension, but it is not just perception. Outdoor spaces that are functional and aesthetically desirable (e.g., paving, weather protection, lighting, landscaping, and furnishings) effectively add livable space. Right-sized homes should always incorporate useful exterior spaces that complement the interior design and regional climate. The key is not to just tack these spaces on but to effectively integrate them into the overall design so they seamlessly flow from interior rooms. Possible exterior spaces include the following:

    - ***Wrap-Around Porches.*** Porches on the west and east elevation provide useful weather-protected space and help control low summer sun.
    - ***Protected Entries/Front Porches.*** It is important to have spaces at the entry door with room for benches or chairs that provide protection from inclement weather and hot summer sun.

Figure 3.7: Annual Sales per Square Foot, 2006.

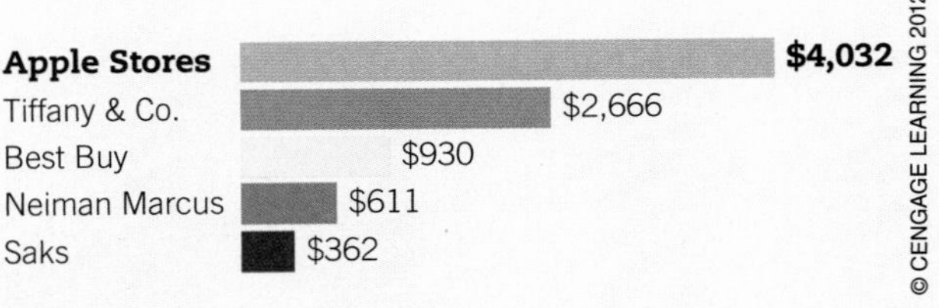

Source: Jerry Useem, "Apple: America's Best Retailer," CNN Money.com, Fortune, March 8, 2007, http://money.cnn.com/magazines/fortune/fortune_archive/2007/03/19/8402321/index.htm

- ***Exterior Screen Porches.*** These spaces are very low cost but can be highly valuable in regions with bugs (most regions). In fact, the screen porch may be one of the favorite rooms in our home because it allows us to sit outdoors during rain storms and cool evenings.
- ***Patios and Decks.*** It is important to have spaces for outdoor cooking, eating, and enjoyment of the property. In some moderate climate regions, outdoor fireplaces are being used to enhance appearance and functionality.

Good Housing Design Criteria 4: Avoid Complexity and Fads

Apple Computer has become one of the most respected technology companies and is often cited as an example of business excellence. Although I defer to much greater industry experts, from my vantage point their winning formula appears to be high-performance products combined with cutting-edge design and continual improvement. Many other industries, including housing, would benefit from this same formula. The concept of high-performance homes is discussed in Chapter 4. Of interest here is Apple's ability to apply outstanding design in every aspect of their business ranging from products to their retail centers. In both cases, the design strategy supplants complexity with elegant clean lines along with quality materials, fit, and details. If sales are any indication, consumers have clearly responded to good design. In 2006 the underground Apple retail store in New York City earned double to more than quadruple the revenue per square foot of other comparable retailers in the city (Figure 3.7). General sales of Apple products and their stock price in a very weak economy are further evidence that superior performance, design, and innovation yield strong business dividends.

In the housing industry, classic design should use windows to capture views, optimize natural comfort, and enhance daylight; select materials that are regionally appropriate and highlight design features; incorporate architectural features that are simple, elegant, and serve the interior spaces; create layouts that provide interest, flow, and needed space for interior furnishings; employ roof lines that effectively drain precipitation away from the home and

Figure 3.8: Southwestern Home with Classic Design.

©iStockphoto.com/Loretta Hostettler.

minimize maintenance; utilize overhangs for shading and protected outdoor spaces (e.g., entries, porches); and conform to natural terrain for ease of access and maximum views. Classic design does not mean that housing architecture has to be boring, but instead means that architecture achieves aesthetic goals by serving design challenges.

Figure 3.8 provides an example of a southwestern U.S. regional design appropriate to very hot and dry climates. It uses simple, regionally indigenous adobe stucco on the outside walls; an appropriate number of windows are placed to capture views and are recessed for shading; and the entry is fully shaded from the sun. Other architectural features include rounded edges to soften the overall appearance, scuppers to drain water, and simple massing that follows the interior layout.

Good Housing Design Criteria 5: Fully Integrate Systems

A wide array of systems in every home should be fully integrated in the final design. This is critical to avoid costs associated with waste, unnecessary work and avoidable service calls, and to maximize opportunities for good design. These systems are discussed in the following sections.

Furniture Layout. Interior and exterior furnishings are not typically considered a system, but homes don't work without "stuff" inside and outside of them. Many potential design problems become evident after the owners move into the home unless furnishing needs are carefully identified and addressed in the design process. This needs to be done in the programming phase with the client or thoughtfully estimated for spec homes. Schematic design layouts should always include likely furniture placements for each room. This will enable optimum placement of walls, windows, doors, HVAC supply and return grills, lighting

Figure 3.9: Calculate Roof Slopes to 2-ft or 4-ft Dimensions.

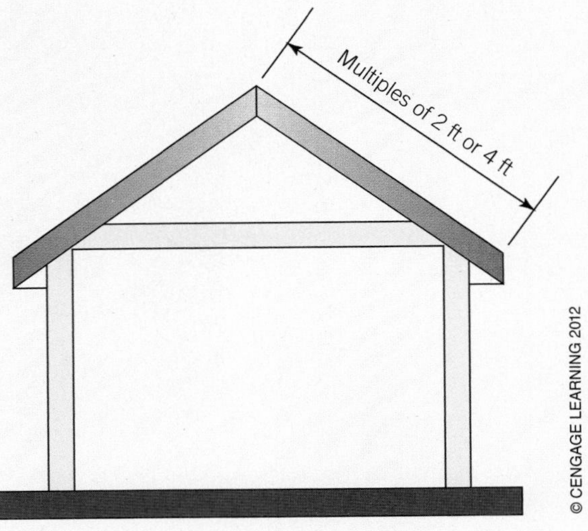

fixtures, electric switches and outlets, cable connections, and built-in cabinets. For example, in bedrooms there often is one obvious place for dressers next to closets that affects the location and size of the closet door.

Structural System. Waste can be significantly reduced by designing homes to standard dimensions used for framing and materials:

- **Board products come in 4 ft dimensions, so floor plans should adhere to 2 ft dimensions to minimize cutting time and material waste.**
- **Framing is commonly spaced 16 in. or 24 in. on center (2 in. × 4 in. studs and 2 in. × 6 in. studs, respectively), so windows and doors should be located on 16 in. or 24 in. framing locations when possible to avoid wasted labor cutting framing and the cost for additional studs.**
- **Roof slopes that result in a 2 ft or 4 ft dimension should be used to minimize wasted materials and time cutting roof sheathing (Figure 3.9). No one will notice if a roof slope is not exactly a whole number of inches rise per 12 inches horizontal dimension (e.g., 4.8/12 instead of 5/12 pitch).**

Water Management. House lot locations and design should consider every aspect of how bulk moisture from rain and snow arrives and leaves the property. This will affect design choices for locating the home on the lot, for windows and entries, and for choosing roofing configurations. Roofs should be configured to completely drain water away from walls, circulation areas, and internal corners. Pockets created where complex roofs allow snow and water to accumulate against walls are asking for trouble and are likely sources of maintenance. Windows should be located where there is generous space for complete flashing details. Grading must ensure water drains completely away from the home.

Disaster Resistance. Every home design should minimize the effects of disasters common to the region. This includes setting home elevations at least 3 ft above base flood elevations in flood-prone areas; considering operable shutters that provide storm protection in regions exposed to hurricanes, tornadoes, and high winds; specifying noncombustible materials with adequate clearances in fire-prone regions; and using stable house geometry in areas exposed to earthquakes.

HVAC System. Forced-air duct systems are the most common form of space heating and cooling distribution used today. However, these systems take more planning than hydronic systems that distribute hot water through pipes to radiators and convectors. First, HVAC ducts should always be inside the conditioned space; no exceptions. Heated air is typically 105° F and should not be distributed through subfreezing attics or crawl spaces. Air conditioned air is typically 55° F and should not be distributed through superheated attics that can exceed 140° F. Tried-and-true design solutions for locating ducts inside conditioned space include dropped ceilings, conditioned attics (e.g., insulating the sloped roof rather than the flat ceiling), and conditioned crawl spaces (e.g., sealing the crawl space, covering the floor with plastic taped at the seams, and insulating the exterior wall rather than the floor).

Another HVAC system integration concern is the location of ducts, registers, and thermostats. Ducts should not be located in exterior walls to avoid unintended heat loss and gain. Supply and return registers and thermostats should be located so they don't conflict with likely furniture placements. In addition, consideration should be given to the location of the exterior air conditioner pad relative to the location of the indoor air handler unit. I have inspected homes in which the air conditioner pad could easily have been relocated at no extra cost to a location that would have saved 25 ft of refrigerant line for the HVAC contractor. I've also visited homes in which the HVAC equipment has been located in the attic where future replacement would require cutting a hole in the roof because the equipment didn't fit through the attic hatch. An adequately sized access should always be provided for future replacement of equipment.

Plumbing System. Plumbing systems can be substantially simplified if fully addressed in the design process. In particular, locating all rooms with plumbing fixtures off one "wet wall" where all the plumbing is located can minimize pipe runs. This is called a "core plumbing" layout. It minimizes wait time for hot water and a corresponding amount of wasted water going down the drain while reducing the cost for installing the plumbing system. Where long runs are unavoidable, advanced technology can come to the rescue in the form of structured pumping systems that minimize wasted water and provide near instant hot water regardless of the home size and layout (see Chapter 5).

Lighting and Electrical System. Effective lighting can completely transform the look and feel of a home and should be carefully designed by a lighting consultant or designer trained in lighting design. The circuit breaker box is typically located in the basement in cold climates and at the garage wall in warmer climates. This does not have significant impact on the design. However, electrical switch and outlet locations need to be considered in design. Again, this can only be done by integrating them with expected furniture layouts in the design process. This will help identify the need and location for potential floor outlets, switches next to beds to shut off overhead lights, cable outlets for entertainment systems, and high-speed Internet outlets convenient to desks and other appropriate locations. In anticipation of rapid market growth for electric cars, all garages should be equipped with car battery charging stations.

Control Systems. Technology for monitoring and controlling all systems in new homes is rapidly expanding, including integration with entertainment, Internet through broadband, whole-house music distribution systems with built-in speakers, security systems, and solar thermal and electric systems. Furthermore, advanced control systems can operate lighting, HVAC, window treatments, appliances, and even irrigation systems. Industry experts are forecasting that all of these systems will be managed by one control device much like the current cable television control, integrated into a computer control center, and accessible from cell phones or dashboard displays, much like in automobiles. Ultimately, homeowners will be able to view their ongoing energy consumption and monitor all their technology systems. Where and how to integrate all of these systems needs to be fully integrated in the design process.

Solar Systems. Public policy along with tax rebates and other incentives have rapidly increased the adoption of solar systems across the country. When homes do not include solar installations in the initial construction, thousands of dollars can be saved on future systems by providing these solar-ready low-cost improvements:[2]

- **Provide adequate south-facing roof (+/– 45 degrees off true south) for both solar thermal and solar electric panels including any required structural load accommodations.**
- **Set aside locations for solar equipment including storage tanks for solar hot water heating and inverters and electric controls for solar electric systems.**
- **Provide at least two extra circuit breakers dedicated to a future solar electric system.**
- **Provide bypass valves on the cold water feed to the hot water tank to accommodate easy connection to a future solar storage tank.**
- **Provide conduit to run piping and wiring from the roof to solar equipment locations.**

[2] "Draft Renewable Energy Ready Home (RERH) PV Construction Specifications," and "Draft Renewable Energy Ready Home (RERH) Solar Thermal Construction Specifications," EPA, October 2010.

PUTTING IT ALL TOGETHER

There are many examples of home designs that effectively demonstrate the principles laid out in this chapter such as those provided in Sarah Susanka's *Not So Big House* series.[3] I've selected two of my own conceptual design projects for discussion here.

Example 1: 1,350 Square Foot Home. When working with the California Energy Commission in 1984, I participated in an advisory team put together by the local utility to develop energy-efficient production housing concepts with a panel of Sacramento area builders. The programming requirements established for the project by the builder panelists were a three-bedroom home with a front entry garage on a prototypical high-density subdivision lot (50 ft wide by 90 ft deep). Figure 3.10 shows the conceptual design I contributed to this project. A detached rear garage off of a back alley would have been preferred to allow for a front porch. This would enhance neighborhood interaction and isolate homes from potential pollutant sources in garages (e.g., residual car exhaust and possible fumes from stored gasoline, paints, fertilizers, solvents, and insecticides). The following sections identify how specific design principles presented earlier were applied even in this 25-year-old project:

- **Fit to Site.** It was assumed there were no compelling views and that the sites were substantially flat, typical of production housing subdivisions in the Sacramento area. With tightly packed houses leaving minimal side yards, the primary view for each lot would be the front streetscape and rear yard designed for living with patios and rich landscaping. Regionally appropriate materials were used including stucco walls for excellent durability under intense Sacramento sun and cement roof tiles, which offer superior durability and fire resistance.

- **Right-Sized Home.** This home provides a good example of how full function can be achieved in a relatively small size following the 10 criteria of right-sized homes discussed earlier. Although 1,350 square feet is very small for a three-bedroom home by today's standards, it features generous spaces: Kitchen with full working island, Great Room with large entertainment center, Master Suite with a walk-in closet and Master Bath, Hall Bath that can be divided with a pocket door to accommodate two users at a time, Bedrooms with large closets, full Laundry Room, and Entry.

  Many features make the home feel bigger including an *open layout* that links the Kitchen, Eating Area, Great Room, and Entry into one large space. *Eliminated/combined rooms* were employed by forgoing the little-used formal dining room and combining the living room and family room into one Great Room. Although this layout option is getting more mainstream today, this was an important innovation when the house was designed 25 years ago. Remaining rooms are accessed with *minimal circulation* space using a small compact hallway that seems bigger with a window providing *natural daylight*. South orientation with full shading provides plentiful *natural daylight*. Spaces also feel bigger

[3] Susanka, *The Not So Big House Book.*

Figure 3.10: Sample Small Home, 1,350 Square Feet.

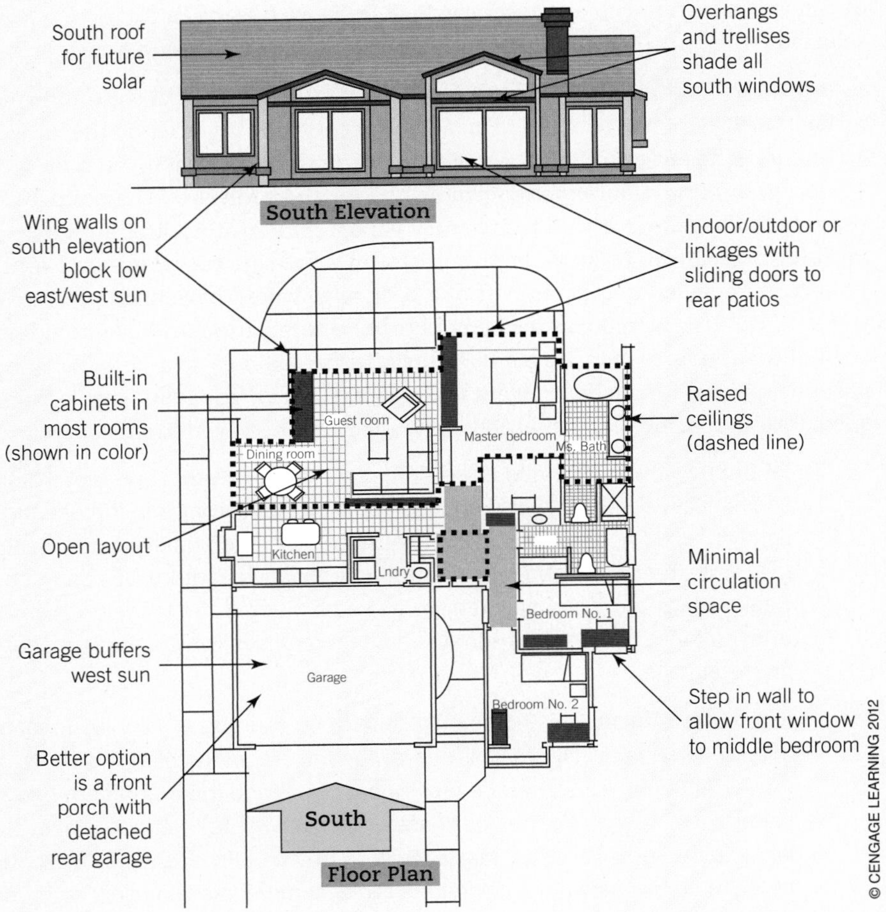

with *varied ceiling heights*. This was easily integrated in the design by choosing structural insulated panels for walls and roofs with raised ceilings following the slopes of the panels. In this home, sloped ceilings were designed for all south-facing rooms and the entry. Large glass doors on the rear elevation adjoining the exterior patio provide *outdoor/indoor* visual linkages that make the interior space live bigger. *Built-in cabinets* are used extensively throughout the house including an entertainment center/storage/shelf unit and a low cabinet unit for separation from circulation in the Great Room, full-height dresser/shelf unit in the Master Bedroom, linen cabinet unit in the Hall, and dressers and desks in the bedrooms. *High quality trim and hardware, advanced lighting design,* and *effective use of designer colors* were also intended to be used throughout the home but do not show up on a floor plan.

- **Integrate Natural Comfort.** This home design shows how natural comfort features can easily be incorporated in traditional housing. It was assumed the rear of the home would face south for this design, so extensive south-facing windows are located on the back elevation. Summer heat gain was the primary concern for this Sacramento, California, location because the summer air conditioning season is very long with many extended periods at or above 100° F (lovingly referred to as "heat storms" in the Sacramento Valley), and the winter heating season is relatively short and mild, often with a thick fog that can reduce the benefits of passive solar heating. Thus all south-facing glass includes appropriate overhangs and trellises that provide full shading in summer and still accommodate winter passive solar heating when available. In addition, wing walls were employed to block angled solar heat gain in the morning and afternoon. The garage is located on the west to buffer the entry and home from the intense, low afternoon/evening sun. In addition, the interior design features extensive thermal mass storage including ceramic tile floors and stone counters and a masonry fireplace wall.
- **Avoid Complexity and Fads.** The design is very simple and functional with simple roof lines and form changes serving architectural functions. For instance, the front elevation is notched in to allow a front-oriented window for the middle bedroom. In other words, the design didn't add a corner for no reason but to avoid a bedroom that only has a window facing a neighbor's home. Additionally, minimal materials are used for walls (stucco) and roof (cement tile), and they are enhanced with effective trim details. Small design features matter. For instance, a raised curb is provided in the garage about 3 ft off the rear wall to stop tires so cars can't accidentally hit the interior wall. This also eliminates a stepped entry from the garage to the laundry room.
- **Integrate All Systems.** Low ceiling areas were utilized over the central part of the house to accommodate useful conditioned space for extra storage and the HVAC system with a compact duct layout. The small footprint allows all plumbing fixtures to be within a short distance of the water heater, which is centrally located in the laundry room. This results in minimal pipe runs and very short duration waiting for hot water. The large expanse of south-facing roof serves as a great location for future solar water heating and solar electric systems that can easily make this a "net zero" home. Net zero means there is no net utility bill; solar system electric production over the course of the year balances out or exceeds annual electric consumption. Finally, dimensions were intended to accommodate standard structural insulated panel sizes.

Example 2: 1,800 Square Foot Home. The second project was also completed over 25 years ago at the California Energy Commission for a report on how to design net zero energy homes. I wish we had had the insight to call this project "Net Zero Homes" at that

Figure 3.11: South Orientation with West-Facing Lot: Floor Plan.

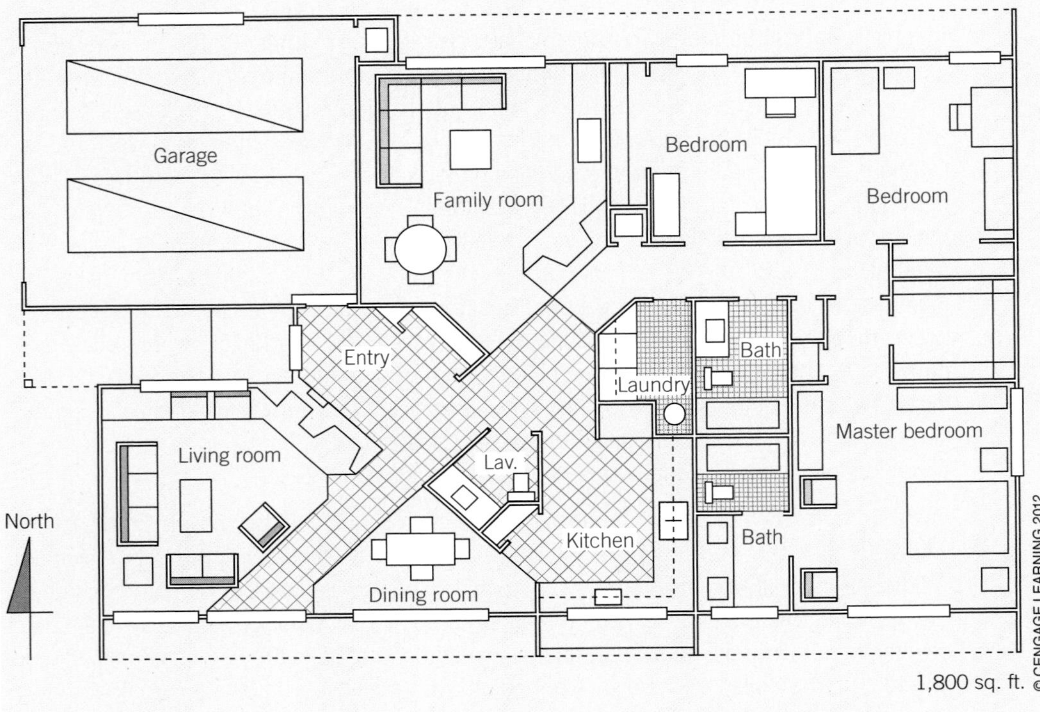

time. Instead we used the awful name "Optimized Cost Low Energy Consuming Photovoltaic Homes."[4] It is no wonder this report sat dead on the shelves for decades when it effectively established concepts stimulating great interest today. As they say, marketing is everything. One home design featured in this report provides a design solution integrating natural comfort in a west-facing lot. The floor plan is shown in Figure 3.11 with a three-dimensional view in Figure 3.12. It utilized a much more traditional floor plan for that time than the previous example. The following sections identify how specific design principles presented earlier were applied to this project:

- **Fit to Site.** A typical flat Sacramento lot without compelling views was assumed for this project, but with a north-south street configuration. This created a significant natural comfort design challenge because the south orientation results in most windows facing

[4] Sam Rashkin and Michael DeAngelis, "Optimized Cost Low Energy Consuming Photovoltaic Homes: A Cost Effective Design Approach for Grid-Connected Photovoltaic Homes," California Energy Commission, June 1984.

Figure 3.12: South Orientation with West-Facing Lot: Three-Dimensional View.

the side yard. This design assumed integration with the land developer during the site planning process to provide wide lots on north-south streets for functional side yards. Additionally, regionally appropriate cementitious panel siding and cement roof tiles were chosen for easy availability, durability, and fire resistance.

- **Right-Sized Home.** This home is significantly smaller than the typical 2,200 square foot size for three bedroom homes, but it features a full traditional layout including separate Family Room and Living Room, formal Dining Room along with spacious bedrooms, bathrooms and storage. Today, I would have combined rooms and created a much more open layout, but there is still some semblance of an *open layout* with visual linkages between the Kitchen, Dining Room, Family Room, Living Room, and Entry. Bedrooms are accessed with *minimum circulation* off the main public space. The south orientation and shading provide plentiful *natural daylight* including the use of a sawtooth roof configuration to provide additional daylight to the Family Room and bedrooms on the north side of the home. The layout also uses a structural insulated panel system that accommodates *varied ceiling heights* in the open spaces and north bedrooms. Large windows and sliding glass doors on the south elevation adjoining the side yard provide *outdoor/indoor* visual linkages that make the interior space live bigger. *Built-in cabinets* are used in the Living Room, Family Room, Kitchen, and Laundry. *High-quality trim and hardware, advanced lighting design, and effective use of designer colors* were also intended to be used throughout the home but do not show up on a floor plan.
- **Integrate Natural Comfort.** Extensive south-facing glass with carefully designed overhangs is included on the south elevation and clearstory windows to deliver desired solar heat gain, shading, and daylighting to both south- and north-facing rooms. Wing walls are used on the south wall and at clearstory windows to block low east and west sun.

Significant thermal mass is provided with tile floors, tile counters, and masonry fireplaces. Both fireplaces are intentionally located on internal walls to minimize heat loss and a cold flue at start-up. A west-facing garage and west wall without windows provide an effective buffer from the difficult low west sun in summer. The resulting home can be expected to have substantial daylight and comfortable mean radiant surface temperatures (warm in winter and cool in summer).

- **Avoid Complexity and Fads.** Clearly this is not a perfect design and is dated in several respects. The front garage in particular dominates the front elevation, but it was assumed a mandatory requirement for typical production builder projects. That said, there is a purpose to every architectural detail used in this design. This includes using a sawtooth roof configuration to allow for high south-facing clearstory windows for north-facing rooms; overhangs designed for proper shading; wing walls designed to block the low west sun from hitting south-facing windows; and an extended front roof overhang to provide a protected porch entry.
- **Integrate All Systems.** Low ceiling areas are included over circulation space, bathrooms, closets, and the Kitchen. This provides needed space for the HVAC system and a compact duct layout as well as additional storage. This design is particularly unique in its ability to locate all wet rooms (Kitchen, bathrooms, and Laundry) except the half bath off a "wet" wall so that all fixtures are within 10 to 12 feet of the water heater. The half bath was intended to have its own in-line electric water heater, which is cheaper and more efficient than running an additional hot water line to this room. This will reduce plumbing needs by 50 percent or more from a typical three-bedroom home and virtually eliminate waiting time for hot water. The three-dimensional view shows that plenty of south-facing roof area was provided for solar electric and solar domestic hot water heating systems that easily can make this a net zero energy consumption home. Finally, dimensions were intended to accommodate standard structural insulated panel sizes.

How Good Housing Design Got Here

Regionally Responsive Designs Prevailed to the End of World War II

Until the 1950s, home designs were commonly responsive to locally prevailing climates and used indigenous building materials. This was critical to deliver some modicum of comfort and to address limits to transporting materials. Examples of regionally responsive home designs follow:

- **Northeast and Midwest.** Comfort during long cold winters was maximized with small window areas to reduce winter heat loss and minimal north-facing windows for protection from prevailing cold winter winds. Fireplaces were centrally located rather than on exterior walls to improve the natural draft and maximize the heat exchange with interior rooms. Front entry vestibules were provided to buffer the interior from cold weather when front doors were opened. Widely available wood was utilized for structure, siding, and trim. Steep pitched roofs were used to effectively shed snow.

- **Southwest.** Comfort during long, intense hot-dry seasons was maximized with large overhangs that blocked the sun, thick adobe or block wall construction that would temper the daytime heat until much cooler evening hours, and small windows to minimize solar heat gain. Stucco walls and tile roofs were utilized for added durability against damaging sunlight and fire protection.
- **Southeast.** Comfort during long, hot-humid summers was maximized with large overhangs that blocked the sun. Concrete block wall construction was commonly used in coastal regions and functioned as an excellent reservoir cladding system to absorb substantial amounts of moisture and provide better resistance to termites and hurricanes. High ceilings, tall windows, and central hallways with doors at each end were used to optimize cross-ventilation cooling with prevailing breezes. Houses were elevated on posts and piers to provide air circulation under the floors and greater flood protection leading to the crawl space foundation. Front porches became a ubiquitous feature of American architecture, providing places to escape the heat and socialize with neighbors.
- **Pacific Northwest.** West of the Cascades, homes experience relatively moderate winter and summer temperatures but extensive overcast and rainy conditions from fall through spring. In response, homes were designed with regionally indigenous wood cladding that was highly resistant to wet weather (e.g., redwood and cedar) and utilized crawl spaces to raise homes above wet ground conditions.

Developments after World War II Changed the Rules

Technology development, economic forces, and government policy combined to profoundly affect mainstream U.S. home design soon after World War II ended. Three critical changes were:

- ***Affordable air conditioning*** meant builders no longer needed climate-responsive designs to ensure comfort. The advent of affordable room and central air conditioning systems starting in 1950 opened up the whole southern portion of the country to year-round comfort heretofore only possible in cold climates.
- ***Cheap energy*** including the expansion of electric, oil, and gas delivery infrastructure spread throughout the country. This meant that increasingly bigger homes could be reasonably affordable to heat and cool without climate-responsive designs.
- ***Explosion of suburban development*** meant production builders were continually getting pushed farther from urban centers in search of large tracts of low-cost land. As a result, they needed compelling reasons to get home buyers to live on more marginal land with longer commutes and less access to urban amenities.

BRUTE FORCE DESIGN REIGNS AFTER WORLD WAR II

The new developments following World War II led to three significant impacts on mainstream home designs:

- **Home designs got bigger**
- **Home designs employed increasing numbers of windows**
- **Home designs were less reliant on regionally responsive designs**

Home Designs Got Bigger. House size has experienced a continuous and steep increase for decades, often without a commensurate increase in function. In particular, builders have enticed American home buyers with increasingly larger kitchens, master suites, bathrooms, family rooms, and storage. In addition to square footage, house size has also increased in terms of volume. Ceiling heights have increased from 8 to 9 ft and often 10 ft along with double-height spaces for entries and special rooms. Double-height ceilings are losing favor because they are hard to maintain and blatantly sacrifice useful square footage, but high ceilings appear to be here to stay.

Putting this all together, from 1950 to 2000 the average house size more than doubled from less than 1,000 to nearly 2,300 sq ft while the number of occupants per household decreased from just under 3.5 to just over 2.5 (Figure 3.13). This means, the average American home between 1950 and 2000 nearly tripled in square footage per person. During this same time, builders have been catering to Americans' love affair with their cars, and the industry transitioned from just offering street parking, to carports, to one-car garages, to two-car garages, and finally to three- and four-car garages in many upscale developments.

This growth in U.S. house size has far surpassed that of the rest of the world. Statistics during the housing boom show that the average size new home in the United States is more than double the average home size in United Kingdom and Sweden and more than 50 percent bigger than the average home in Japan (Figure 3.14).

Home Designs Employed Increasing Window Area. Although exact data is hard to pin down, personal observations suggest window-to-floor area in many regional markets nearly doubled from about 10 to 12 percent in 1950 to about 20 percent at the peak of the housing boom in 2006. This was particularly evident in many high-growth southern states where it was not uncommon to observe window-to-floor area above 25 percent. This trend was reinforced with the introduction of advanced low-e windows in the mid-1980s, which significantly reduced the risk of comfort and condensation problems. However, even high-performance windows only have an R-value of about 3. Combined with increased framing

Figure 3.13: The Growth of House Size, 1950 to 2000.

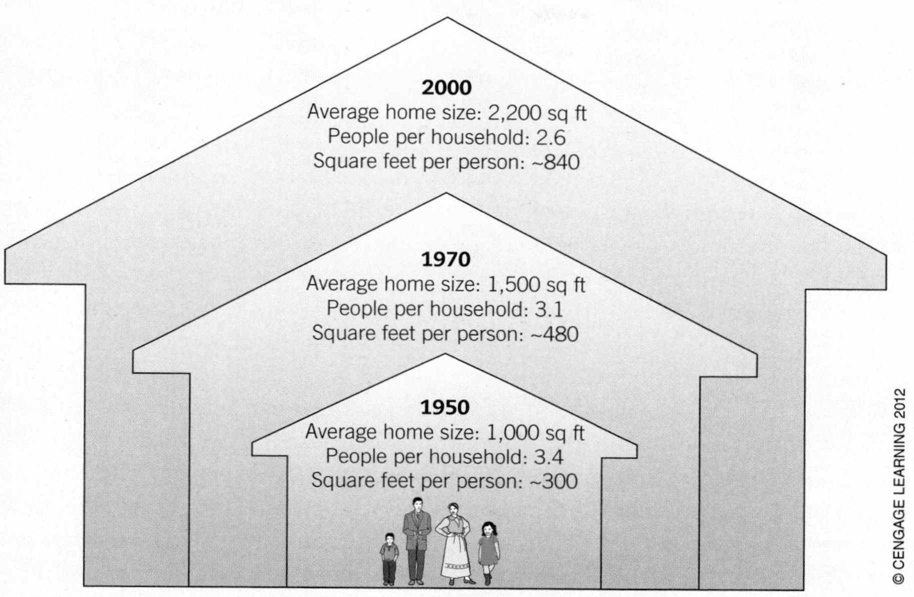

Source: Rick Diamon, Mithra Moezzi, "Changing Trends: A Brief History of the U.S. Household Consumption of Energy, Water, Food, Beverages and Tobacco," August 22, 2004, Lawrence Berkeley National Laboratory, http://epb.lbl.gov/homepages/rick_diamond/LBNL55011-trends.pdf

Figure 3.14: Home Size Comparison for Developed Countries.

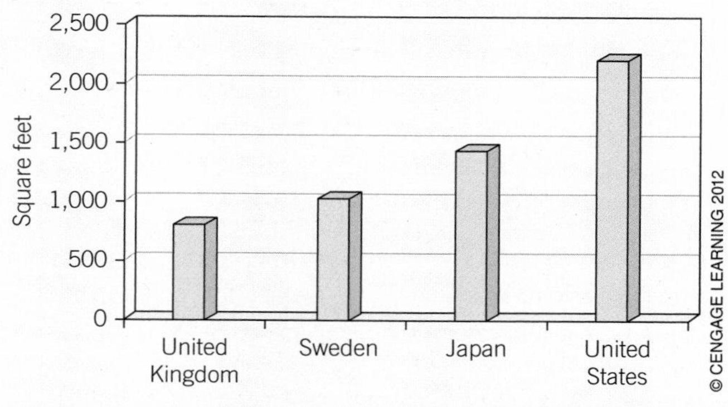

Source: Demographia, International House Sizes, 2002, www.demographia.com/db-intlhouse.htm

factors (term for percent of wall area comprised of wood studs, headers, and plates) that often exceeded 25 percent (see Chapter 4), about half the wall area in many homes effectively had little insulation.

Home Designs Became Less Regionally Responsive. With "brute force" heating and air conditioning systems, home builders no longer employed climate-responsive designs to maintain comfort, even as the average home size kept getting bigger. Cheap energy prices reinforced this design trend, with home utility bills representing only a small percent of total expenses for the average household. Homes were also decoupled from their regional environment as the use of indigenous design solutions and materials declined. This was partly in response to the use of stock plans employed by large production builders across divisions in different parts of the country. This maximized their investment in developing those designs and helped reduce their construction costs. Moreover, many locally available products became scarce (e.g., wood) or expensive (e.g., brick) and were increasingly replaced by low-cost manufactured materials. Both aluminum siding, introduced in the 1960s, and vinyl siding, introduced in the 1980s, were embraced as cheap alternatives to actual wood siding. Vinyl siding has now substantially supplanted aluminum siding and has become the leading cladding option in many midwestern and eastern regions of the United States. As brick became more expensive, the design solution was to limit its use to the front elevation. This became known as "facade architecture." And as car ownership exploded, so did the need for garages that came to dominate the front elevations of homes. Between garages and air conditioning, the font porch soon disappeared as a prominent architectural feature. Since World War II, homes have become much more homogeneous and less regionally responsive across large geographic areas of the United States.

Recent Trends Will Impact Housing Design

Statistics and demographics don't lie, and they are signaling profound changes to come in the world of housing. The following trends need to be embraced in new home design:

- ***Smaller Homes.*** For the first time since the housing industry began keeping records, in 2009 the average home size has gotten smaller. The recent first-time home buyer tax credit has reinforced that trend in 2010. Although some industry experts do not forecast a long-term size reduction trend, there are a whole host of reasons why it should continue: reduced family size, projected baby boomer movement to smaller homes, latest Gen X and Gen Y buyers' preferences for smaller homes, and broad interest in homes with lower energy and maintenance costs.

- ***"Greener" Homes.*** The jury is out on how the green movement will evolve, but most surveys indicate consumers are looking for more environmentally responsive solutions. "Green" is a market force that will have legs if advocates can navigate their way through the consumer backlash to inevitable "green washing"—over promising and sometimes unscrupulous claims. This is especially true if the housing industry effectively puts forward the compelling value proposition with green homes: lower cost, more comfort, improved indoor air quality, greater durability, and a more sustainable future for our children.

- ***More Technology-Savvy Homes.*** The Gen X and Gen Y population will soon be the largest new home buyer demographic. To be sure, this generation brings a whole new set of design preferences. In addition to their strong preference for right-sized homes, they love advanced technology. There will be a definite correlation between this demographic group's interest in buying new homes and the degree technology is effectively integrated in new designs.

Why Good Housing Design Is Broken

A substantial improvement in customer satisfaction would be realized if the housing industry applied the criteria for good design presented in this chapter:

- **Feature better fit to site to consistently capture views, experience fewer drainage problems, include front porches that enhance social connections, and apply regionally appropriate materials for greater visual appeal and durability**
- **Incorporate natural comfort with heating, cooling, and daylighting that feel better and cost less**
- **Offer fully functional spaces with less wasted space, more storage, less clutter, and richer materials, trim, and accessories**
- **Feature classic design with architecture that supports important functional improvements and less complexity for fewer maintenance problems**
- **Fully integrate all systems so they cost less to build, have fewer defects, and feature advanced controls for optimal performance**

Instead, personal observation suggests that the housing industry has made what appears to be a Faustian deal, being willing to lure new home buyers with cosmetic "look and feel" and size over designs that stand the test of time. These failures to apply the five good housing criteria previously presented are discussed next.

Homes Are Not Consistently Fit to Site

Too often production builders place standard plans on sites without consistently optimizing available views, drainage solutions, use of front porches, or regionally specific designs that can maximize comfort and energy efficiency. Additionally, standard home plans used across numerous regional divisions dilute attention to regionally responsive designs that effectively address local climate conditions and indigenous materials. A few examples follow:

- ***Wood Shake Roofs in California.*** When I moved to California in 1981, wood shake roofs were ubiquitous even though they are a serious fire risk and have poor durability under the intense sunlight in the western region. This incredibly flagrant violation of regionally appropriate design came to a head during the Oakland Hills fire in 1991 as evidenced by a prominent newspaper image showing only one home standing among the barren ruins.

This was the only home that had a clay tile roof while the other homes that completely burned to the ground had wood shake roofs. Although wood shake roofs have since declined in use, the willingness of the California housing industry to make a combustible material the roof of choice in an extremely fire-prone region exposed a very broken design process.

- ***Exterior Insulation Finish Systems (EIFS) in the Mid-Atlantic.*** EIFS systems have an elastomeric stucco finish that can trap moisture inside walls where installed without a drainage layer. Drainage layers were commonly not included. This is not a good thing in hot-humid climates. Prominent moisture failures ensued in many states but were especially egregious in heavy rainfall states like North Carolina and Virginia (see Chapter 5). Although it's possible to install EIFS systems that work in these climates, EIFS failures remain another testament to the risks of using regionally inappropriate materials.
- ***Air Conditioning Air Handler and Ducts in Hot Climate Attics.*** There is no sugar coating the fact that the housing industry is horribly broken when it comes to locating space conditioning systems in hostile attic conditions. It represents a complete rejection of regionally appropriate design. Attics in hot desert climates can reach 160° F and higher. That is not a good environment for distributing 55° F conditioned air. This design strategy also results in a significant number of ceiling penetrations for supply and return grills that compromise the thermal boundary between home and attic. In gulf and southeastern states, attics are also incredibly humid, contributing to significant moisture control challenges, and dust, pollen, and pests can easily infiltrate leaky ducts and air handlers in attics to further compromise comfort and indoor air quality. The housing industry has embraced this epidemic failure for the sake of first-cost savings. However, homeowners are paying the price in terms of significantly higher energy costs, reduced comfort, and poor indoor air quality.

Natural Comfort Is Not Integrated

Modern heating and cooling systems have allowed builders to employ designs that boldly ignore prevailing temperature, humidity, and wind conditions. As a result, production builder homes feature substantially increased window areas with no attention to solar orientation and overhangs for proper shading. These principles of natural comfort have been known for decades, most commonly under the name "passive solar design." Often the opportunity for natural comfort is lost in the development process because builders do not put enough pressure on subdivision planners to optimize the number of north-south-oriented lots. This is an especially significant lost opportunity on sites not constrained by varied terrain and compelling vistas.

Homes Are Not Right-Sized

Unfortunately, size became a "lazy" metric to ratchet up consumer demand for new homes further removed from urban centers. As homes became increasingly bloated, customer satisfaction was often short-lived due to design flaws such as excessive wasted space, poor accommodation to furniture layouts, and cleaning and lightbulb replacement issues in double-height ceilings.

The housing industry recently experienced its first reduction in average new home size since World War II, with a decrease from 2,473 sq ft in 2008 to 2,422 sq ft in 2009.[5] And this reversal occurred before the impact of the first-time home buyer tax credit that stimulated sales of smaller sized homes. However, many experts are not forecasting a long-term shift to right-sized homes. Moreover, the housing industry still does not employ many of the principles of right-sized homes: built-in furniture to increase effective useful space, high-quality trim and hardware to enrich interiors, effective use of color, daylighting, and expert lighting design.

Faced with significantly more volume and surface area in larger homes, it appears many builders were forced to control hard costs by using cheaper materials and finishes. I have observed high-end, supersized homes with various combinations of inexpensive windows, doors, trim (baseboard, window, and door), hardware, cabinets, lighting fixtures, switches, outlets, and plumbing fixtures. In the world of production homes, baseline low-cost finishes are often camouflaged with highly decorated and upgraded models that function like a "bait-and-switch" sales proposition. Once the infatuation with newness is over, homeowners often realize their homes are equipped with bland colors, cheap materials, and the most basic lighting fixtures. This cannot be a good design strategy for ensuring customer satisfaction that fuels long-term growth.

Complexity and Fads Too Often Prevail

Frequent changes to cosmetic finishes and tacked on architectural components suggest that the housing industry relies on trendy design features to mainline "emotion" into the new home sales process at the expense of classic style that will stand the test of time. This includes jumping quickly to the latest trends with new cabinet materials, finishes, and countertops for kitchens and bathrooms, new appliance finishes, and construction details like rounded drywall corner beads. Many large production builders use design centers to showcase these latest cosmetic offerings in settings that effectively contrast their exciting possibilities with the drab kitchens, bathrooms, and finishes in prospective buyers' current homes. Obviously this emotion-driven marketing has proven highly effective, but it is short-sighted, without integrity to the rest of the design process.

One of the most egregious examples of "fake" design popular in many parts of the country is a concept called facade architecture. As discussed earlier, the most desirable exterior finishes (brick, stone, and wood) and architectural interest (pop-out walls, bay windows, dormers, and special trim details) are employed only on the front elevation; the other three elevations are finished with demonstrably lower-cost materials such as vinyl siding and designed with straight flat walls. This is a blatant attempt to control costs, especially for larger homes. Moreover, it's just plain silly because it effectively assumes owners and passersby only see the home in two dimensions straight on from the front and won't notice the awkward transitions between high-end and cheap finishes (Figure 3.15). Taken to its worst extreme, facade

[5] Alan J. Heavans, "U.S. Average House Size Shrank in 2009," Philadelphia Inquirer, June 18, 2010, Builder Online, http://builderonline.com/economic-development/us-average-house-size-shrank-in-2009.aspx

Figure 3.15: Facade Architecture with Awkward Transition.

Figure 3.16: Facade Architecture with Small Side Return Wall.

architecture has even been applied to short return walls (Figure 3.16), making them look cheap while yielding very small cost savings. Homes look better when all four elevations are integrated into a complete design.

Of course it is possible that I am an architectural snob. But I have to believe even untrained homeowners subliminally recognize the complete absence of a holistic approach to the design process, particularly when it screams out at you. And facade architecture effectively does scream out, functioning like a giant billboard declaring to all passersby, *"the front of our home is what we would like the whole house to look like, but could not afford."* Yet in many markets consumers have been beaten down by the proliferation of facade architecture and the lack of a better alternative.

Unfortunately, there are many examples of housing design gone wrong. Here are a few of the most egregious errors:

- **Substantial amounts of glass with no purpose, no coordination with likely furniture placement, or lack of consideration for sun control issues**
- **Excessive variation in cladding materials**
- **Awkward floor plans and room shapes that don't optimize useful space and likely furniture placement**
- **Dormers and ornate roof lines with no purpose and complex details difficult to construct with proper moisture protection (Figure 3.17)**
- **Excessive bump-out walls that complicate flashing details and drainage**
- **Designs that do not provide adequate weather protection such as covered entries, or inadequate clearances for complete window flashing**

Design Does Not Integrate with All Systems

The housing design process pays only minimal attention to the diverse systems in a home. As a result, different trades are left to make suboptimal decisions in the field as they address a variety of design constraints:

- **For decades the housing industry has blatantly ignored the fact that sheet products come in 4 ft × 8 ft standard sizes, and framing follows set 16 in. and 24 in. on center spacing. When determining roof slopes or locating windows and doors, these dimensions are often ignored, resulting in significant wasted wood and framing time.**
- **HVAC systems often are not integrated into designs so they can be installed with minimal time, cost, and defects. Instead, poorly trained installers configure ducts in the field without proper sizing; install the ducts with excessive lengths, too tight bends, compression in tight spaces, and inadequate support that severely restricts air flow; and locate supply and return grills in suboptimal locations.**

Figure 3.17: Home with Complex Roof Design.

- Heating and cooling equipment and ducts are commonly located in unconditioned attics and crawl spaces, which results in about 10 times more heat gain in summer and heat loss in winter than ducts placed inside the conditioned space. In addition, a large number of penetrations between the conditioned home and unconditioned spaces are created that compromise the thermal boundary of the home. This is a substantially deficient design flaw in home construction.
- Lighting appears to be one of the housing industry's lowest priorities. Most builders miss a major opportunity to enhance the ambiance and perceived size of their homes with effective lighting.
- The housing industry is ill-prepared to address the technology and automation preferences for the largest wave of new Gen X and Gen Y home buyers. Their comfort with and expectations for technology are substantially greater than the baby boomer generation that has dominated the industry's home buying demographic for decades. New technology to consider includes control and monitoring

systems for broadband entertainment and information systems, whole-house music distribution with built-in speakers, HVAC systems, lighting, solar systems, and security systems.

How to Fix Good Housing Design

Suggesting significant changes to how the housing industry designs new homes challenges what many builders consider their core competency. But it's time for the housing industry to reject old design methods that rely on brute force comfort, complicated architecture, increased size, and tacked on cosmetics. Dressing up model homes with over-the-top decorating and highly marked-up upgrades is no longer good enough to drive sales (see Chapter 6). Buyers know good design when they see it, and good design can create a compelling reason to invest in a new home. The following sections discuss some important retooling recommendations for the housing industry.

Invest in Design Experts

I hope there is general agreement that design is critical to the housing industry. Then it shouldn't be so hard to invest meaningful resources into the best designers, and not just architects. Design experts should be considered for the following critical components:

- ***Architects*** who are experts on preferred design styles as well as bulk water management, natural comfort, right-sized homes, and comprehensive systems integration.
- ***Landscaping specialists*** who can specify indigenous and drought tolerant species of ground cover, shrubs and trees, and who can integrate porches, patios, fencing, trellises, and decks effectively with the home architecture.
- ***Interior design specialists*** who can select rich color schemes, specify quality trim and hardware, and integrate built-in cabinets throughout the home.
- ***Lighting experts*** who can design appropriate lighting for each room and specify the latest high-efficiency and solid-state fixtures, switches, outlets, and controls.
- ***Plumbing experts*** who know how to design efficient plumbing layouts that use less piping, deliver water to fixtures quickly with less waste, and specify water efficient fixtures.
- ***Technology experts*** who can incorporate entertainment, information, security, and home automation into new housing.

Investing in this expertise will result in a whole new level of fit, finish, and functionality that will take new housing to the next level.

Return to Housing's Regionally Responsive Roots

Natural comfort should be maximized for lots with, and without, a north-south orientation. In other words, efforts should be made to develop custom and model home designs that optimize south-facing glass and properly designed overhangs for different compass orientations. In addition to reduced heating and cooling costs, natural daylight will be enhanced by allowing occupants to maintain year-round comfort and views during daylight hours without covering the windows. Heavy window treatments would not be needed in the summer to block undesirable sunlight, and in winter window treatments would be left intentionally open to increase desirable solar heat.

Regionally responsive design should also be employed to provide architectural interest on all four elevations, using appropriate materials. This includes integrating front porches and other weather-protected outdoor spaces, using window areas for maximum effectiveness, and roof designs that maximize water protection.

Right Size the Right Way

The industry needs to get over the crutch of increased size to attract buyers. There are signs that change is starting to happen. The bigger challenge is to right size effectively. After applying the principles of right-sized homes discussed in this chapter, homes will live 20 to 30 percent bigger. In other words, builders will be able to construct a 2,000 sq ft home that lives with the full function and spaciousness of a 2,500 sq ft home. Right-sizing principles addressing open layouts, eliminating or combining rooms, minimized circulation, varying ceiling heights, and daylighting do not require additional construction costs. A reduction of 500 sq ft at $80 per square foot of construction costs results in about $40,000 lower first cost. What should be done with these cost savings? The knee-jerk reaction would be to pocket them. Times are hard and the market is soft. This is a temptation the industry must reject. Instead, it's critical that the savings be invested in the quality improvements addressed in the right-sizing principles:

- **Provide built-in furniture throughout the home**
- **Use high-quality trim, hardware, and finishes**
- **Offer several coordinated color palettes chosen by a color expert**
- **Integrate high-quality lighting throughout every room**
- **Visually link indoor spaces with desirable outdoor spaces**

Americans have embraced dramatically smaller cars than decades ago because of impressive improvements in design; ergonomics; interior fit, finish, and trim; performance; safety; and fuel economy. The same formula will work for homes. It's time to start engineering the next generation of American right-sized homes by investing the savings in square footage into design, quality details, and performance (see Chapter 4).

Avoid Unnecessary Complexity and Fads

This is a call for the housing industry to embrace four-sided architectural designs that have integrity and purpose. It's time to stop bad habits like these:

- **Slapping on fake fronts**
- **Maximizing complexity of roof geometries while ignoring practical details for water protection and drainage**
- **Indulging in excessive different materials that compete with each other rather than enhance the architecture and regionally appropriate design**
- **Throwing in lots of angles that ignore feasible furniture layouts**
- **Adding excessive window area with no real purpose and a negative impact on energy bills and comfort**
- **Building double-height spaces that are hard to maintain, light, often appear out of place, and sacrifice a significant amount of otherwise useful space**

If you can afford only a limited amount of brick or stone, don't use it all on a large front elevation; use the same amount of material for the base finish on all four elevations (e.g., up to a few feet above grade) or to highlight a special architectural feature such as a prominent pop-out section or large chimney. If you want an interesting roof, use more classic configurations that provide proven water protection with higher-quality roofing. Instead of using lots of different materials just for the sake of variety, use a stronger design strategy enhanced by regionally appropriate and durable materials. If you want more interesting floor plans, use angles selectively to enhance flow and create visual linkages in open layouts. If you want windows to provide maximum impact, use them to enhance natural comfort (heating, cooling, and lighting), maximize views, and to highlight architectural features. If you want to create visual interest and a sense of space, use varied ceiling heights rather than double-height ceilings.

Fully Integrate All Systems

It's time for home designers and architects to recognize that there are a wide array of systems that affect floor plans, elevations, and spatial volumes. For example, only 12 in. truss-joist floor framing might be needed but an increase to 14 in. truss-joists might provide adequate clearance to run HVAC ducts between floors inside the conditioned space rather than in an unconditioned attic. This type of HVAC system consideration is commonly absent from the design process; this has to end. The housing design process must engage other experts and trades as necessary to ensure optimal choices are made that result in better quality at lower cost. Specifically, the following systems must be integrated for compelling improvements in new home designs:

- **Furniture layouts** need to be included on floor plan designs to ensure optimal flow; adequate size for each room; and proper locations for windows, doors, lighting fixtures, switches, outlets, HVAC supply grills, entertainment cable outlets, and broadband connections in each room.
- **Structural systems** choices must be coordinated with HVAC and other systems to ensure adequate space requirements as well as to minimize waste and cost. Consider framing spacing and sheet good dimensions before setting floor plan dimensions, locations of windows and doors, and roof pitches.
- **Water management systems** that drain water from roofs, walls, openings, site, and foundation need to be carefully considered in all final design choices. Visual interest is never an excuse for adding risk of moisture problems.
- **Disaster resistance systems** should be considered in all design choices to minimize the risk of damage from weather, natural events, and pests.
- **HVAC system** should not be an afterthought. It bears repeating that it is time to stop the madness and locate all HVAC systems inside the conditioned space. There are several design options for fixing this problem. The first is to use unvented attics that are insulated at the slope rather than at the flat ceiling. This effectively converts the attic to conditioned space and creates design possibilities for adding storage or potential loft space. Alternatively, ducts could be run in dropped ceilings while maintaining insulation at the flat ceiling above. Where dropped ceilings are used for ducts, they should be coordinated with right-sized home design strategies employing varied ceiling heights. It may be possible to run ducts between floors with truss-joist and open web framing. This will be an increasingly feasible option as homes get much more efficient (see Chapter 4) and need much smaller ducts with more compact layouts. HVAC supply and return grills should be coordinated with expected furniture placement, and HVAC equipment locations should be coordinated with the design to minimize undesired noise and to ensure access for maintenance and future replacement.

- **Plumbing systems** need to be given much more attention to save costs and improve performance with much quicker hot water service at each fixture. In some cases, core layouts with all wet rooms located off one wet wall will be possible. Where core layouts are not possible, the design should include structured pumping systems (see Chapter 5). Both of these plumbing design options will substantially reduce the amount of wasted water down the drain waiting for hot water and provide near instant hot water at each fixture.

- **Lighting and electrical systems** should be fully addressed with a dedicated lighting plan for each floor that lays out task, ambient, and mood lighting solutions for each room. These plans should include specifications for appropriate bulbs and fixtures; locations of fixtures, switches, and outlets; coordination with architectural lighting details such as cove lighting, indirect lighting from above cabinets, and spot lighting for art niches and displays; and control systems.

- **Solar systems** are becoming increasingly cost effective and attracting more homeowner interest. If not part of the initial design, thousands of dollars in future installation costs could be saved by adding simple, low-cost details. All homes except those in heavily forested areas or with very limited solar resources should be solar ready. Designs should ensure adequate roof area with south orientation (+/- 45 degrees of true south) and structural capabilities for collectors; dedicated pathway and conduit to run wires and pipes; locations for additional solar hot water tank and inverter; shut-off valves on cold water pipe to the water heater to allow easy installation of a future solar hot water storage tank; and extra circuit breakers.

- **Control systems** are rapidly emerging for new homes that will provide a quantum jump in homeowners' ability to interface with the diverse systems in their homes. I often characterize this development as the "i-Home," and anticipate these homes will have great appeal to the next wave of Gen X and Gen Y home buyers. These homes will include digital displays that monitor home systems and can be accessed on a home computer or through remote operation devices such as cell phones. Homeowners will be able to control a variety of systems including lighting, HVAC, entertainment, security, irrigation, window treatments, and whole-house music. In addition, the display provides real-time feedback on energy use and production of solar systems where installed. It is time for housing design to embrace this new technology.

Chapter 3 Review

So What's the Story?

Design is so personal and abstract a concept, it is with profound humility that I have provided detailed guidelines in this book. But it is critical to get design right in all industries, especially housing. The good housing design story can be summarized as follows.

- **What It Is.** Good housing design ensures that a home addresses its local setting, spatial needs, design integrity, and wide array of systems.
- **How It Got Here.** Regionally responsive design strategies gave way to brute force design with the advent of cheap energy and air conditioning, but recent trends demand changes to this model.
- **Why It's Broken.** New home designs ignore critical opportunities to improve quality, comfort, and durability.
- **How To Fix It.** Invest in experts who can effectively develop right-sized home designs with improved quality and natural comfort fully integrated with all systems.

The details are included in Table 3.2.

Table 3.2: Good Housing Design

What It Is	How It Got Here	Why It's Broken	How To Fix It
Good housing design criteria: 1. Fit to site: Prevailing views Drainage patterns Livable community Region 2. Integrate natural comfort 3. Right-size: Open layouts Eliminate or combine rooms Minimal but generous circulation Varied ceiling heights Built-in furniture High-quality trim, hardware, and finishes Effective use of colors Daylighting Effective lighting design Indoor-outdoor linkages 4. Avoid complexity and fads 5. Fully integrate systems: Furniture Structural Water management Disaster resistance HVAC Plumbing Lighting and electrical Control Solar	Regionally responsive designs prevailed until the end of WWII Developments after WWII changed the rules: • Affordable air conditioning • Cheap energy • Suburban development Brute force design reigns after WWII: • Bigger size • Increasing amount of window area • Less regionally responsive Recent trends will affect housing design: • Smaller homes • "Greener" homes • More technology-savvy homes	Homes do not meet good housing design criteria: 1. Homes not consistently fit to site 2. Natural comfort is not integrated 3. Homes are not right-sized 4. Complexity and fads too often prevail 5. Design is not integrated with systems	Invest in design experts. Return to the regionally responsive design roots. Right size designs the right way. Avoid complexity and fads. Fully integrate all house systems.

Retooling THE U.S. Housing Industry

4

High-Performance Homes: Why Homes Work and Fail

High-Performance Homes: Process, Goals, and How Goals Are Achieved

Process. *The concept of a "high-performance home" is relatively new to the housing industry. It entails three fields of expertise that emerged once homes began to incorporate meaningful levels of insulation in the 1980s and after the insurance industry recognized that the frequency and magnitude of natural disasters were increasing. These fields include:*

- ***Building science,*** *which examines how to effectively control air flow, thermal flow, and moisture flow in buildings according to basic laws of physics and climate-specific factors*
- ***Indoor air quality,*** *which applies technologies and materials needed to effectively control sources of pollutants, dilute remaining pollutants, and filter particulates inside homes*
- ***Disaster resistance,*** *which identifies the most regionally prominent risks associated with weather, natural events, and pests and then protects homes with appropriate construction improvements*

High-Performance Homes

Goals
- Affordability
- Comfort
- Health
- Durability

How
- Building science
- Effective components
- Pollutant control
- Disaster resistance

© iStockphoto.com/Elenathewise

© iStockphoto.com/Card76

All of these disciplines are applied to high-performance homes during the key steps in the design and construction process. These steps include:

- ***Design schematics***
- ***Final designs***
- ***Contract documents***
- ***Construction supervision***
- ***Construction inspection and testing***

Goals. *For the past quarter century, advocates for high-performance homes have been relentlessly promoting four essential goals for high-performance homes. The first is greater affordability based on ownership costs rather than on first cost. Ownership costs include the ongoing costs to pay the mortgage, utility bills, additional medical expenses caused by indoor air quality problems, and ongoing maintenance for a home. The second goal is to provide even room-by-room and floor-by-floor comfort throughout a home. The third goal is a healthy indoor environment. This is essential for good home performance because each day people spend 90 percent of their time indoors, including about 60 percent of their time spent inside the home.[1] Finally, a high-performance home should be durable, with a full array of construction details and practices that ensure long life with minimal maintenance.*

How. *The basic laws of physics dictate that high-performance goals are directly linked to comprehensive building science. This is because air, thermal, and moisture flow measures deliver:*

- ***Increased affordability with lower energy bills resulting from reduced heating and cooling loads, and lower maintenance costs resulting from less risk of moisture damage***

[1] U.S. EPA, "The Inside Story: A Guide to Indoor Air Quality," U.S. EPA/Office of Air and Radiation. Office of Radiation and Indoor Air (6609J), cosponsored with the Consumer Product Safety Commission, EPA 402-K-93-007, April 1995, http://www.epa.gov/iaq/pubs/insidest.html

- ***Improved room-by-room and floor-by-floor comfort with an effective thermal enclosure that minimizes heat loss and gain, controls surface temperatures, and reduces noise***
- ***A healthier indoor environment with moisture control measures that help avoid mold and dust mites due to excessive humidity, and air flow measures that limit the entry of dust, pollen, and pests***
- ***Improved durability with tight construction that reduces condensation in wall assemblies, comprehensive water protection that can eliminate moisture damage, and high-performance windows that reduce UV degradation to finishes and furnishings***

Efficient components enhance the affordability of high-performance homes by minimizing the energy required for heating and cooling loads, appliances, lighting, and other plug loads.

Pollutant control is achieved with three key strategies that minimize the risk of indoor air quality problems:

- ***Source control with products and construction practices that minimize or eliminate dangerous chemicals, moisture, radon, pests, and combustion by-products***
- ***Dilution of remaining pollutants with whole-house ventilation systems that provide fresh air, spot ventilation systems that exhaust moisture and pollutants from bathrooms and kitchens, and garage ventilation that exhausts a wide array of potential pollutants resulting from car exhaust and stored materials***
- ***Filtration that uses effective filters in the heating and cooling system to remove particulates and dust***

Disaster resistance is achieved through specialized construction practices and materials that address risks relevant to each location, including extreme weather (e.g., floods, blizzards, hurricanes, and tornadoes), natural events (e.g., fires and earthquakes), and pests (e.g., termites).

What Is High Performance?

Four Components of High-Performance Homes

A wide array of experts are likely to offer various interpretations of what constitutes the basic framework for high-performance homes. Based on extensive relationships with many of these experts and experience promoting energy-efficient construction to thousands of builders, I have identified four key components that define truly high-performance homes:

- **Building science**
- **Energy-efficient components**

- **Pollutant control**
- **Disaster resistance**

Each of these components is critical to consistently achieve the goals of affordability, comfort, health, and durability.

High Performance Starts with Building Science

There are more than 130 million existing single-family and multifamily homes in the United States. More than 95 percent of them evidence poor performance based on commonly occurring symptoms: unnecessarily high utility bills; seasonal discomfort; cold and hot floors over garages and cantilevers; poorly managed interior moisture levels that can lead to musty smells, mold, and dust mites; epidemic levels of respiratory health problems; widespread thermal defects; high indoor noise levels; window condensation; excessive bugs; high levels of dust; ice dams in winter; and uneven peeling paint. All of these symptoms can be attributed to the absence of comprehensive building science. Like the mythical frog that will stay in a gradually heated pot of water until it dies, American homeowners have come to accept poor performance as the norm. But this will change once a critical mass of new home buyers experiences the substantial advantages of high-performance homes and realizes they come at lower ownership cost.

The Rules of Building Science

I have identified 10 critical rules of building science that explain why buildings work and fail (funny how lists always add up to 10). Homes that effectively apply these rules will raise performance to dramatically new levels.

Building Science Rule No. 1

Air, Thermal, and Moisture Flow Always Move in the Same Direction

It bears constant repetition that the reason homes work or fail has everything to do with controlling three driving forces that are both naturally occurring and man-made: air flow, thermal flow, and moisture flow (Figure 4.1). This is the basic underpinning to building science.

These driving forces act from both inside and outside of homes as shown in Figure 4.2. Air flow (green arrows) includes wind; radon, which is radioactive air below grade commonly found in granitic soils; stack effect, which is hot air rising; and induced air flow by an assortment of fans inside homes. Over the last few decades, many fans have been integrated in homes, creating greater challenges for controlling air, thermal, and moisture flow. Potential building failures associated with these fans are addressed later in the chapter. Thermal flow (orange arrows) includes exterior heat gain in hot weather and interior heat loss in cold weather. Heat inside the home is generated intentionally from space conditioning systems, fireplaces, and wood stoves, and unintentionally from internal loads including human activity, cooking, and waste heat from equipment. Moisture flow (blue arrows) comes in two forms: vapor, which is commonly referred

Figure 4.1: Three Key Principles Drive Building Science.

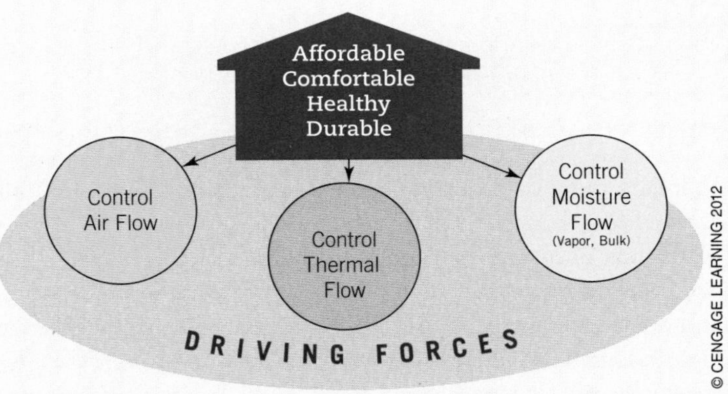

to as humidity; and bulk, which is water. Moisture vapor flow occurs due to exterior humid weather and interior humidity from breathing, washing, cooking, cleaning, and humidification. Bulk moisture occurs as precipitation (e.g., rain or snow) and as ground water.

If building science is all about controlling these driving forces, the good news is we know their direction of flow. The Second Law of Thermodynamics taught in high school physics dictates that air, thermal, and moisture always move from more to less heat, pressure, and humidity. Once you identify the source of more heat, pressure, or humidity, you know the direction of flow. This is critical to effective air sealing, insulation, and placement of air and vapor barriers.

Figure 4.2: Driving Forces Acting on Homes.

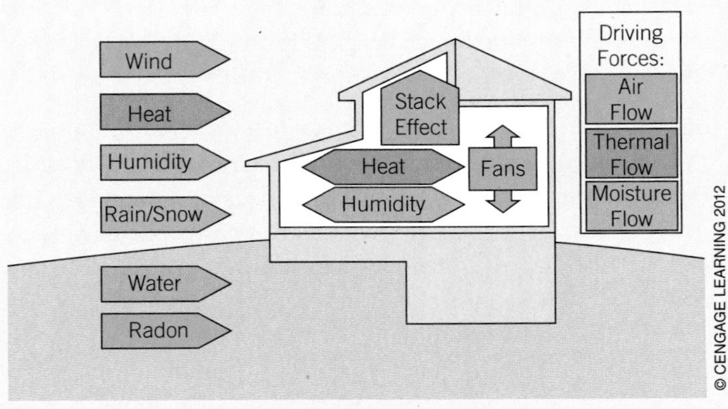

Building Science Rule No. 2

All Action Happens at Cold Surfaces

Where air, thermal, and moisture flow move from "more to less" and have an opportunity to hit a cold surface below the dew point (surface temperature at which moisture condenses out of air), condensation will occur. As long as the rate of wetting is less than the wall system's capacity to dry, this will not cause a moisture related problem. All construction materials get wet to some degree. However, when materials get wet beyond their ability to store moisture and dry, this can create a building failure that leads to mold and rot damage.

The risks associated with condensation in construction assemblies today are much greater than with older homes for two reasons. First, wood framing is less tolerant of drying. Lumber has much higher moisture content, and widely used engineered wood products (e.g., plywood, oriented strand board) are composed of compressed and glued layers or particles that deform and lose their structural integrity much more quickly than solid wood. Second, homes are better sealed and insulated, a trend that will continue with scheduled code changes. Tighter construction and more insulation creates colder surfaces, which substantially slow the rate of drying and create greater opportunities for condensation. Insulation and comprehensive air sealing were rarely used in pre-1960s construction. This allowed internal heat to flow easily to the outside sheathing and keep it above the dew point (Figure 4.3). Thus old buildings are durable, but the trade-off is that they also are more costly to heat and cool, and can be very uncomfortable.

The introduction of insulation and air sealing resulted in construction assemblies where there is much greater resistance to thermal flow. In contrast to older homes that weren't insulated, surface temperatures below the dew point occur much more easily at the exterior sheathing during cold weather. This can lead to cold weather condensation problems when air flow has a path to the exterior sheathing. Figure 4.4 shows one way this can happen in walls with poorly sealed electrical outlets and switches.

Figure 4.3: Walls Without Insulation Allow Heat Flow.

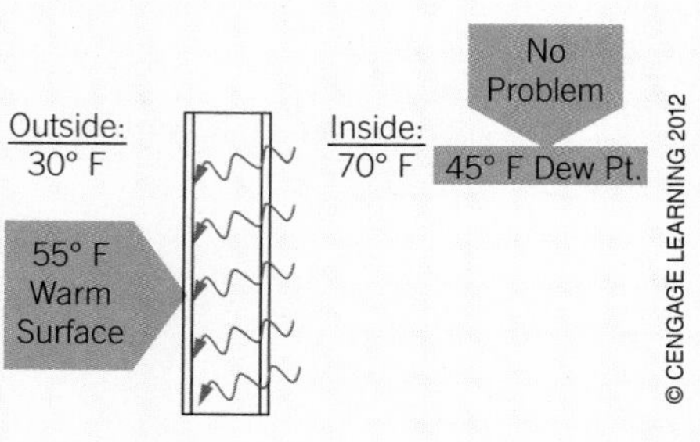

Figure 4.4: Insulated Walls Create Cold Surfaces.

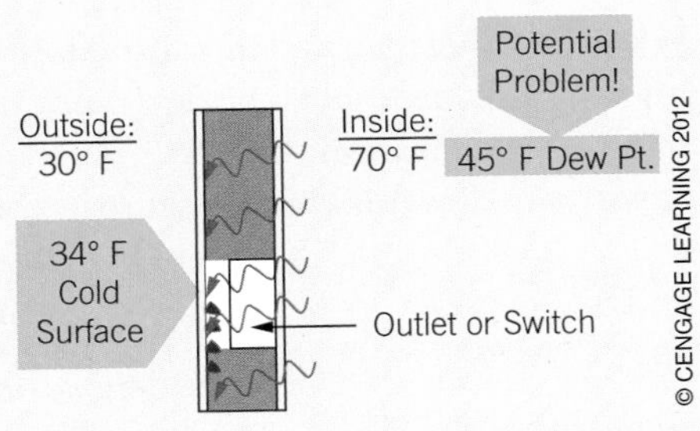

Figure 4.5 shows a home in which a leaky electrical outlet provided a path for air and moisture vapor flow as diagrammed in Figure 4.4 that resulted in significant mold and rot damage. Does this happen every time there is an air leakage defect? No; this happens under conditions when the condensation at surfaces exceeds the ability for the wall assembly to dry. ***Controlling air leakage is critical in insulated homes.***

Building Science Rule No. 3

There Are Two Conditions for Air Leakage: Holes and Driving Forces

No matter how significant driving forces may be, air leakage does not occur without holes. Think of the extreme exterior pressures and temperatures acting on an airplane flying at 30,000 feet elevation: winds in excess of 100 mph and temperatures below –40° F. Comfort can be maintained inside an aircraft under these most extraordinary conditions only by controlling the holes with virtually air-tight construction. Alternatively, if there is a large gaping

Figure 4.5: Uncontrolled Air and Moisture Flow.

Figure 4.6: Big Holes for Controlling Air Leakage.

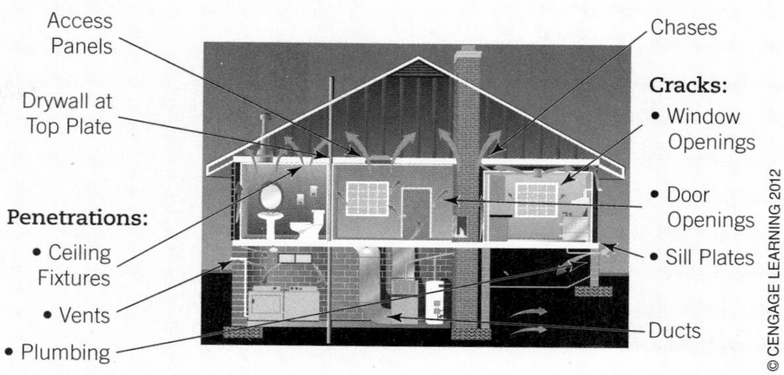

Source: "Air Sealing and Insulation That Works," EPA ENERGY STAR for Homes, December 23, 2010, www.energystar.govindex.cfm?c=behind_the_walls.btw-airsealing.

hole in a building, but absolutely no difference between indoor and outdoor pressure, temperature, and humidity, there would be no air leakage. However, this will never happen; driving forces across construction assemblies are virtually assured. And there is nothing we can do about them because we can't control the weather. *If controlling air leakage is the problem, sealing holes is the solution.* Air takes the path of least resistance, so the most effective air sealing strategy is to seal the big holes first. Fortunately, this is a manageable goal because we know the location of the big holes (Figure 4.6). These big holes are grouped into six categories:

Chases. All chases, also referred to as shafts, must be fully flashed and sealed at boundaries between conditioned and unconditioned spaces such as attics, basements, and crawl spaces. Chases are commonly used for fireplace chimneys, HVAC ducts, flues, plumbing, and other mechanical or electrical penetrations.

Cracks. Gaps at window and door openings and sill plates in an average size home account for about a half-mile of cracks that need to be air sealed.

Penetrations. Any plumbing, ducting, or electrical work that penetrates the construction assembly to an unconditioned space or exterior should be fully sealed. This includes hose bibs; vents for kitchen and bath exhaust fans, clothes dryers, and central vacuums; and all electrical wiring and ceiling light fixtures.

Access Panels. Three types of attic access openings commonly occur in homes: ceiling access panels, drop-down stairs, and attic knee-wall doors. They represent large thermal holes that should be fully insulated and gasketed.

Drywall at Top Plates. Hidden gaps between drywall and wall framing top plates can create extensive thermal bypasses where the top of any interior or exterior wall adjoins an attic above. However, the significance of this leakage problem has only recently been understood. This gap should be fully sealed at all interior and exterior walls with options such as latex foam between the sheetrock and top plate or spray foam along the top of all exterior and interior walls.

Duct Leakage. Ducts are effectively the "lungs" of the home, distributing conditioned and filtered air to each room. They should be fully sealed with highly durable materials to significantly reduce utility bills; ensure room-by-room comfort; protect indoor air quality by drawing in humidity, dust, pollen, and bugs at leaky return ducts; and avoid durability problems by allowing warm air to escape leaky supply ducts in unconditioned attics and crawl spaces where it can reach cold surfaces during winter.

Building Science Rule No. 4

One Out Equals One In

Simply stated, systems are self-balancing. If interior air leaks to the outside (exfiltration) or is exhausted (e.g., ventilation fans), a home will not keep losing air until it goes into a vacuum. Any indoor air lost through leakage or exhausted to the outdoors will be made up by an equal amount of outdoor air being drawn in through the remaining holes and cracks. Similarly, where exterior air leaks indoors (infiltration), the home will not continually inflate and burst like a balloon. Any outdoor air infiltration will be accompanied by an equal amount of indoor exfiltration.

Building Science Rule No. 5

Thermal Control Must Address All Forms of Heat Transfer

High school physics teaches us that there are three types of heat transfer: conduction, convection, and radiation. They each play a critical role in affecting thermal enclosure heat loss and heat gain.

Conduction is heat transfer through materials as molecules exposed to heat excite each other. Materials have different inherent conduction characteristics. Home performance is maximized using materials in construction assemblies with less conduction (lower U-value) or more resistance (higher R-value). For instance, metal is 300 times more conductive than wood. As a result, steel framing functions as a heat transfer superhighway conducting heat loss and heat gain that substantially undermine the thermal performance of the whole assembly. Steel framed walls can reduce the whole-wall effective R-value of an R–13 insulated wall to less than R–6. Therefore, steel framing should only be used with exterior rigid insulation sheathing to provide a complete thermal break against conductive heat loss and gain. Common 2 in. by 4 in. wood studs have an R-value of 3.5, which is nearly four times less than the R–13 insulation that would fit in the wall cavity. This is significant because conventional wood framing can make up as much as one-fourth or more of the walls in a typical home.

There are several options to reduce thermal bridging in walls. One option is to minimize framing to what is needed for structural purposes. This maximizes cavity space available for wall insulation and minimizes wasted cost for excessive wood framing. This is accomplished with a technique commonly known as optimized value engineered (OVE) framing that can reduce framing to as low as 12 percent or less of the wall. Figure 4.7 shows there is much more open cavity space for insulation with the advanced wall framing on the top photo compared to the conventionally framed wall shown in the bottom photo.

Figure 4.7: Advanced Wall Framing.

(a) Advanced framing

(b) Conventional framing

COURTESY OF WWW.BUILDINGSCIENCE.COM

Another option to minimize wood framing is to use structural insulated panels (SIPs). SIPs are constructed in a plant by making a sandwich of wood sheathing (typically oriented strand board commonly known as OSB) and foam insulation (typically expanded polystyrene, but occasionally polyisocyanurate). Wood framing is used inside an SIP assembly only for splines, plates, headers, and rough openings. Depending on the system selected, this framing comprises 3 to 8 percent of the wall. Complete thermal break construction assemblies with no continuous framing are also available. They include wrapping all framed walls with rigid insulation board or using advanced wall systems such as insulated concrete forms and double-wall framing.

Convection is heat transfer resulting from the natural movement of air from more to less heat, humidity, and pressure and from hot air rising (stack effect). The first line of defense is air-tight construction (see Rule No. 3). In addition, insulation assemblies have to be installed without air spaces between the insulation and air barriers (e.g., sheathing, drywall) that enable convective air flow around the intended thermal resistance of the insulation. This is called thermal bypass. Four common insulation installation defects create air spaces: gaps, voids, compression, and misalignment (Figure 4.8). *Gaps* are spaces between the edge of insulation and one of the framing members. *Voids* are pockets where insulation does not completely fill a framing space. *Compression* occurs when insulation is squeezed so that full thickness is not maintained (e.g., insulation force fitted around wiring and pipes or crammed into a cavity). In addition to creating air space for convective flow, compression eliminates air pockets substantially responsible for the thermal performance of insulation. *Misalignment* results when insulation is not in direct contact with an adjoining air barrier.

Figure 4.8: Common Insulation Installation Defects Creating Air Spaces.

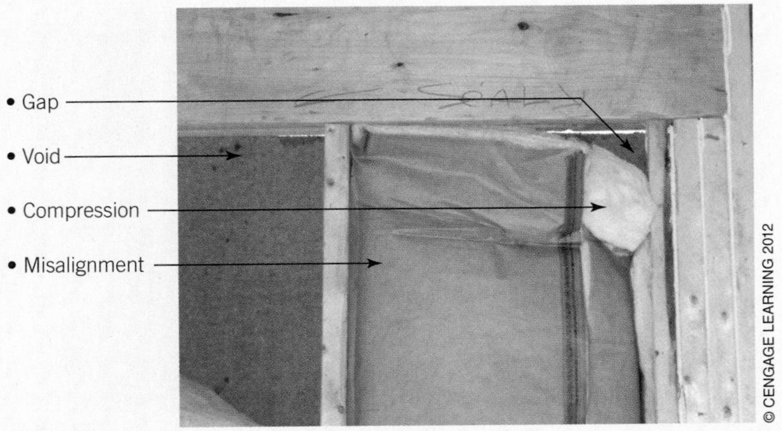

- Gap
- Void
- Compression
- Misalignment

Controlling convection requires complete air barriers in addition to proper installation of the insulation and air sealing. How air barriers are integrated depends on which of the following type of insulation is being utilized:

- **Batt (fiberglass and rock wool)**
- **Blown-in (cellulose and fiberglass)**
- **Board (expanded polystyrene, extruded polystyrene, and polyisocyanurate)**
- **Spray foam (open- and closed-cell polyurethane, with new medium-density products also available)**

Each insulation type has associated costs and benefits. Their relevance to air barriers is that the two most common and low-cost options—batt and blown-in insulation—are typically made of fibrous material that does not completely block air flow. The effectiveness of insulation is a function of a material's ability to be engineered and fabricated with a maximum number of air pockets because contained air has excellent resistance to thermal flow. The air pockets in fibrous insulation are porous enough that air can flow through the material. This means a complete air barrier is critical with fibrous insulation to stop convective flow through the material that could bypass the thermal resistance. An air barrier is any material that blocks air flow such as thin-board sheathing, plywood, OSB, or sheetrock. Note that edges and seams of air barriers must also be fully sealed. Rigid insulation board and open-cell spray foam insulation have much tighter, fully contained air pockets providing a two-for-one deal: insulation and an air barrier. Closed-cell spray foam products are so dense that they provide a four-for-one deal, adding a vapor control function and additional racking strength as well.

Figure 4.9 illustrates how the absence of air barriers with fibrous insulation contributes to thermal bypass. In this diagram attic insulation is laid over a framed soffit without any sheathing material below leaving no bottom-side air barrier. This is a common defect even in new construction.

In the winter condition shown it is freezing outdoors. The attic would experience roughly the same cold temperature as outdoors while the house is heated to 70° F. Drywall has little thermal resistance value, and the space in the dropped ceiling should quickly assume the 70° F room temperature (not shown in the diagram). This results in a driving force of more heat in the dropped ceiling air space to less heat in the cold attic. Since fibrous insulation does not stop air flow, the driving force would result in a more to less flow through the fibrous insulation. However, the "one out equals one in" rule implies there would be a convective loop with cold air flow returning from the attic to the dropped ceiling space. This would happen until some balance point is reached, which is shown in the diagram as 50° F. This creates a cold surface temperature at the drywall affecting occupant comfort. Also note that if fibrous insulation is used in the walls without an integral air barrier, the same driving force could result in air flow through the wall insulation to the cold exterior sheathing. Because all actions happen at cold surfaces, there could be a risk of condensation and moisture damage.

Figure 4.9: Thermal Bypass at Dropped Ceiling.

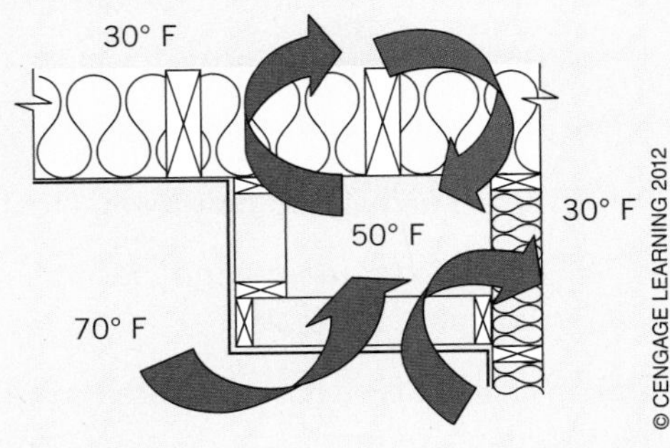

Minor defects leading to convective flow can significantly degrade the performance of insulation. For example, Figure 4.10 shows how effective R-value degrades by 30 to 50 percent with a relatively small percent of missing insulation. All thermal enclosure details for controlling convection are important: sealing holes, cracks, and penetrations; installing insulation with no gaps, voids, compression, or misalignment; and ensuring complete air barriers.

Radiation heat transfer occurs when heat waves travel through space from hotter to colder objects. Most construction assemblies focus on controlling conduction and convection with air sealing and proper insulation installation practices, but there are times where controlling radiation heat transfer is critical to thermal performance. For example, windows cannot use traditional insulation materials and maintain transparency. Advanced aerogel insulation may eventually provide an effective transparent insulation, but the technology is still undergoing

Figure 4.10: Effect of Missing Insulation on Insulation Performance.

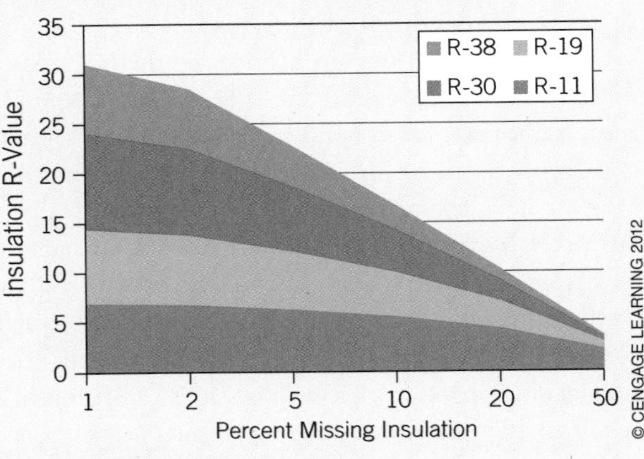

Source: Personal communication with Bruce Harley, December 22, 2010.

Figure 4.11: Radiant Heat Transfer Control with Low-e Windows.

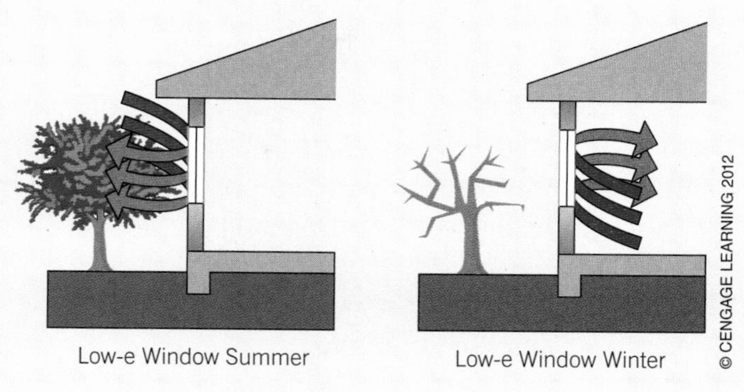

research and development and remains extremely costly. Thus windows rely on radiant barrier technology to maximize thermal performance with advanced low-emissivity (low-e) coatings. These low-e windows employ invisibly thin metal or metallic oxide layers sputtered onto one of the inside surfaces of the dualglazed window assembly. This coating effectively blocks undesired heat gain in summer and heat loss in winter with only a small reduction in visual transmittance (Figure 4.11). This results in better control of the interior glass surface temperature for superior comfort. On average, low-e windows are about 12° F warmer in winter and cooler in summer.

Roofs in hot climates provide another opportunity for improved performance by controlling radiant heat transfer. This can be accomplished with reflective roofing materials or radiant barriers on the underside of the roof sheathing. An increasing array of roofing materials is available today with high reflectance and these materials can be easily identified by looking for the ENERGY STAR® label for cold roofs. Radiant barriers also can be installed underneath the roof sheathing with foils or coatings, or much more simply by purchasing roof sheathing that has a laminated reflective foil directly applied by the manufacturer to the underside. Radiant barriers can drop the temperature of attics up to 30° F and are especially effective when heating and cooling ducts and HVAC air handlers are located in unconditioned attics. However, HVAC ducts and air handlers should never be located in the attic in new home construction (see Rule No. 10).

Building Science Rule No. 6

Mean Radiant Temperature Dominates Comfort

One of the consequences of poor control of air and thermal flow is that surface temperatures get excessively hot or cold. Not only does this contribute to high utility bills, potential moisture problems, and reduced indoor air quality, it also compromises comfort because the temperature of surrounding surfaces substantially dominates occupants' perception of comfort. It is estimated that the mean radiant temperature (MRT) has 40 percent more impact on comfort than the surrounding ambient temperature.[2] For example, we can feel comfortable outdoors on a cold day with the sun providing radiant warmth, but immediately feel freezing cold when out of the sun and next to a cold surface even though the ambient temperature has not changed. Similarly, when we lose control of surface temperatures in homes, we have lost the comfort battle. This leads to raising or lowering thermostat set-points to accommodate one uncomfortable room at the expense of others and creating possibilities for other moisture and high utility bill problems.

Building Science Rule No. 7

Bulk Moisture Must Be Drained

All claddings get wet, no exceptions. Water gets behind wood, brick, stone, stucco, cementitious siding, vinyl, and aluminum, so walls must be effectively drained. In addition, roofs, foundations, and sites must be fully drained and openings flashed for a complete water management system (Figure 4.12).

Figure 4.12: Bulk Moisture Must Be Drained.

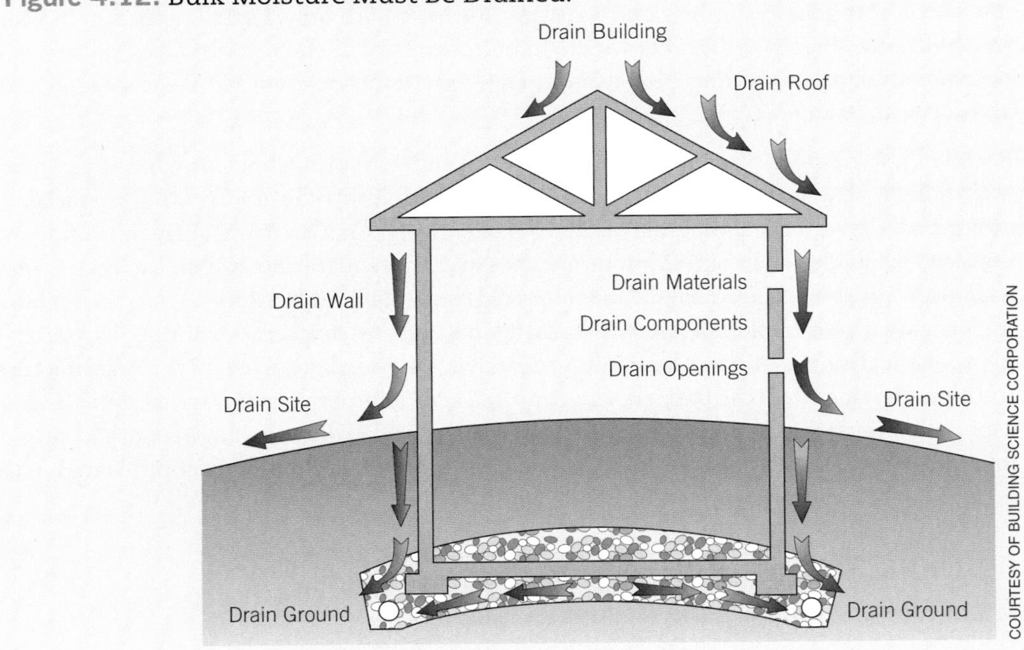

[2] David Wright, *Natural Solar Architecture: a Passive Primer* (New York: Van Nostrand Reinhold, 1978), 19.

It is critical to protect high-performance homes from bulk moisture because they are much less tolerant to getting wet due to tighter construction and properly installed insulation. This is easily understood with the analogy of a hair dryer. Hot air blown directly on wet hair dries it quickly and easily. Place a piece of batt insulation between the hair dryer and your hair and the process takes much longer. Wrap that batt insulation in a plastic bag to simulate a complete air barrier assembly, and drying your hair will appear to take forever. The same is happening when construction assemblies get wet in high-performance homes. Thus a comprehensive water management system must be provided with every high-performance home and include the following:

Roofs

- **Heavy membranes at valleys, eaves, and penetrations**
- **Drip-edge flashing at all exposed roof sheathing**
- **Kick-out flashing where roofs meet walls**

Walls

- **Complete weather resistant barriers (house wrap, building paper, or exterior insulation taped at all joints)**
- **Pan flashing at all windows and doors (fabric or molded plastic)**
- **Flashing at all deck penetrations**
- **Flashing at bottom of cladding (weep holes, weep screeds)**

Foundation

- **Capillary break at slab (plastic liner over gravel)**
- **Capillary break between foundation and stem wall (plastic, damp-proofing)**
- **Drain tile wrapped with fabric filter**

Site

- **Slope site away from home**
- **Slope patios, walks, and driveways away from home**
- **Gutters and downspouts carry water away from home or to swales**

Building Science Rule No. 8

All Construction Gets Wet; Let It Dry

All construction assemblies get wet to some degree, even high-performance homes that are properly drained and well sealed. This includes moisture flow from diffusion; air leakage through hairline cracks and small remaining defects; bulk moisture penetration during extreme weather events; lack of owner maintenance, such as not fixing plumbing leaks; and detrimental occupant behavior, such as excessive use of humidifiers. So it's important to let homes dry.

Based on Building Science Rule No. 1, the drying process goes from more to less heat. Vapor barriers (any material that blocks moisture vapor flow) should typically not be placed on the cold side of a construction assembly because they will stop moisture flow (a.k.a., natural drying process) dead in its tracks. That's their job. This can lead to condensation with increased risk of mold and rot failures as moisture accumulates. This rule results in different regional recommendations for the placement of vapor barriers. In climates with significant air conditioning loads, vapor barriers should not be placed on the inside wall where they can block drying from outside to inside. This includes regions with very cold winters that also experience hot and humid summers such as Chicago and Minneapolis. Similarly, in climates with significant heating loads, vapor barriers should not be placed on the exterior to prohibit drying from inside to outside. And in mixed climates with both heating and air conditioning loads, vapor barriers should not be placed on either side of the construction assembly because drying needs to occur in both directions.

Building Science Rule No. 9

Build It Tight, Ventilate It Right, and Keep It Safe

Once unintended holes are comprehensively sealed for draft-free construction, intended holes are needed for several reasons: fresh air, exhausting odors and humidity, providing combustion air, and exhausting combustion gases. High-performance homes can experience about one-fourth or less of the natural infiltration found in older less-efficient homes. As a result, adequate fresh air must be supplied to dilute and remove indoor pollutants. The primary pollutants include chemicals in building products (e.g., VOCs, formaldehyde); moisture generated from construction materials drying (e.g., wood and concrete) and from human activity (e.g., cooking, cleaning, breathing, washing); biological contaminants (e.g., pests, dust mites); radon; combustion by-products (e.g., cooking, candles, smoking); and pollutants introduced by occupants (e.g., chemically treated furniture, deodorizers, and all sorts of nasty stuff carried in on the soles of shoes).

Whole-house mechanical ventilation systems provide intended holes to bring in and to exhaust air. Typically, they operate continuously, but some use controls to operate for a set amount of time each hour. Systems that only exhaust air, not surprisingly, are called "exhaust-only" systems. They work on the "one out equals one in" rule: exhausting indoor air induces an equal amount of outdoor air infiltration through the remaining cracks and holes in the construction. This creates a slight negative pressure, so this type of ventilation is not recommended in hot/humid climates to minimize the risk of condensation at cold interior drywall surfaces.

Systems that only supply fresh air are called "supply-only" systems (this is starting to make sense). Typically these systems connect a small duct from an outdoor vent into the HVAC return air duct ahead of the filter rack. When the HVAC air handler fan is operating, it will also pull air in from the outdoor supply air duct. With incoming air entering ahead of the air handler, it is filtered and conditioned (cooled and dehumidified in hot weather and heated in cold weather) before being distributed through the home's duct system. This also works on the "one in equals one out" rule: outdoor air flow drawn in induces an equal amount of exfiltration through the remaining cracks and holes in the construction. Because this creates a slight positive pressure, this type of ventilation can be a concern in cold climates where the induced exfiltration of warm indoor air could reach cold exterior sheathing surfaces and condensate. However, recent research suggests that the amount of induced positive pressure is not significant enough to cause a problem.

Finally, whole-house ventilation systems that simultaneously supply and exhaust air are called "balanced" systems. As a result of balanced flow, homes are neither under positive or negative pressure, making these systems appropriate for all climates. They ventilate more effectively than supply- or exhaust-only systems by controlling both intake and exhaust air flow. However, they can be much more expensive. The most common types of balanced systems are heat recovery ventilators (HRVs) and enthalpy recovery ventilators (ERVs). HRVs improve efficiency because they typically recover 70 percent or more of heat in outgoing air in the incoming air in winter or transfer 70 percent of incoming heat in air in the outgoing air in summer. Balanced systems with ERVs include both heat recovery and moisture exchange for additional efficiency and comfort. Incoming hot-humid air transfers a portion of its heat and humidity to the outgoing air in summer (e.g., enhanced control of relative humidity in summer), and outgoing air transfers a portion of its heat and relatively higher humidity to the incoming cold dry air in winter (e.g., avoids drying out the home in winter). Balanced systems can be implemented at lower cost by simply combining an exhaust-only and supply-only system.

Many builders and consumers question the prudence of working so hard to seal holes only to create new ones for whole-house ventilation. The reason for this strategy is control. First, unfiltered air coming in through accidental holes can introduce moisture, dust, pollen, and pests. In addition, relatively humid air flow through unintended holes has the opportunity to hit cold surfaces and cause moisture and mold problems. Whole-house ventilation systems properly installed also maximize the quality of air entering the home by locating intake vents safely away from pollution sources such as chimneys, driveways, and exhaust vents. In addition, the air coming through unintended holes is unreliable and may provide excessive or inadequate dilution to maintain indoor air quality.

A number of other intended holes are also needed to exhaust localized pollutants. This includes spot ventilation fans in baths and kitchens to exhaust excessive moisture, odors, and by-products of cooking through ducts to outdoors (e.g., exhaust vent at walls, soffits, or roofs). Clothes dryers include exhaust vents that must be ducted to the outside to remove moisture from the drying process. Note that ventless clothes dryers are discussed later, which condense the moisture out of the air and do not need a vent. However, they are primarily available in Europe and not yet widely accessible in the United States. Central vacuums pull substantial amounts of dusty air through a central canister that also must be ducted to the outside, and combustion appliances and fireplaces use exhaust flues or pipes to remove combustion by-products.

Natural draft exhaust furnaces, boilers, and water heaters require a fire-rated exhaust flue and outdoor combustion air intake ducts when located in a conditioned room. This compromises comfort because the intake air duct is effectively a large open uncontrolled hole. Additionally, there are safety concerns where negative pressures inside the home can lead to back-drafting of combustion gases or flame roll-out under severe conditions (e.g., very leaky return duct or prolonged use of cook-top fan). As a result, high-performance homes should always use direct-vent furnaces, boilers, and water heaters to ensure safety. This equipment completely isolates the combustion process from the indoor environment by providing separate plastic pipes for intake of outdoor combustion air and exhaust of combusted air. There are also significant utility bill savings because direct-vent equipment has higher energy-efficiency ratings than naturally drafted equipment and eliminates additional heating and cooling loads caused by the large outdoor combustion air intake ducts. Similarly, high-performance homes use fireplaces with outdoor combustion air intakes for safety and efficiency advantages.

Building Science Rule No. 10

Install HVAC Systems Properly to Control Air and Moisture Flow

HVAC systems must be installed properly to achieve fully rated equipment efficiency, ensure proper air flow to all rooms, optimize humidity control, and maximize equipment life. Proper installation begins with right-sizing equipment, ducts, and terminals. This entails working with engineered calculations in accordance with established industry standards based on construction specifications, number of occupants, and regional climate conditions. Part of the right-sizing process should also ensure that the sensible heat ratio of specified equipment does not exceed the level calculated so it can address the latent load. If it doesn't, the home must be equipped with a central dehumidification system to effectively manage relative humidity.

Proper sizing of HVAC systems typically results in smaller equipment, especially for high-performance homes that have very small heating and cooling loads. Similar to the way the car industry used to promote increased horse power for enhanced performance, historically more tons of cooling or BTUs of heating have been promoted by the housing industry for better performance. The opposite is true. Here's why.

Oversizing leads to short-cycle operation as load requirements are quickly met. This can result in humidity control problems during the cooling season when the heat exchanger coils do not get cold long enough to effectively remove moisture. In these cases, occupants often lower the set-point temperature to control moisture levels but compromise thermal comfort with colder than desired ambient temperatures. During the heating season, short cycling leads to excessive "cold blow" at the start of each cycle because air that sits in ducts in unconditioned attics and crawl spaces gets cold between longer periods of operation. All of these problems also increase costs. First, initial cost is greater for larger equipment. Short cycling also imparts more wear and tear on equipment, much like a car operating in stop-and-go traffic rather than at continuous highway speeds. This increases maintenance cost and ultimately replacement costs because of reduced lifetime of the equipment. Lastly, operational efficiency is lower, which increases energy costs.

Some of the oversizing problems are reduced or eliminated by using variable speed or dual-speed equipment with high part-load efficiencies. However, the HVAC industry associations still strongly recommend right-sizing to ensure optimum efficiency and reduce first cost. In an effort to battle misinformation about better performance with larger equipment, I often tell consumers to think of the outdoor air conditioning equipment as an eraser. The bigger it is, the more mistakes the builder has made and is attempting to fix with brute force rather than quality construction.

In addition to equipment, ducts need to be right-sized. In high-performance homes with small heating and cooling loads, full comfort can be maintained with much smaller sized ducts and less extensive duct layouts. Proper duct sizing results in sequentially reduced trunk duct size as conditioned air is supplied to the rooms throughout the home and properly sized branch ducts that are based on the air flow needed for each room or group of rooms served. Ducts in high-performance homes do not need to follow the common practice of providing long runs to outlying locations, typically above or below windows. This was necessary with older homes to counteract drafts from poorly sealed construction and to minimize condensation due to the cold surface temperatures of inefficient windows. Smaller ducts with compact layouts provide another opportunity for builders to reduce first costs while improving performance.

The last component for right-sizing addresses terminals (supply registers). The very low heating and cooling loads associated with high-performance homes also result in very low air flow rates. This demands special attention to the selection of supply registers that are able to fully disperse conditioned air in each room.

In all cases, high-performance homes should locate the HVAC air handler and ducts inside conditioned spaces. In addition to mitigating most of the energy penalty from duct leakage, this substantially reduces conductive energy loss and gain. Heated air is about 105° F, but can be exposed to below freezing temperatures during the winter where ducts are located in attics, crawl spaces, or garages. Air conditioned air is about 55° F, but can be exposed to temperatures over 140° F during the summer if ducts are located in the attic. Considering that most ducts in unconditioned spaces include only R-5 to R-8 insulation, the heat loss and gain is approximately 10 times greater than if located in conditioned spaces. This is because house temperatures are consistently about 70° F, which is a significantly more efficient environment to distribute conditioned air.

Duct leakage is another unintended hole that should be tightly sealed, even when ducts are located inside the conditioned space. This is to ensure that adequate heating and cooling is delivered to each room as intended and to minimize air flow and pressure imbalances inside construction assemblies that can contribute to warm air reaching a cold surface.

Once ducts are sealed and properly sized, their performance can still be undermined with poor installation practices. For example, proper flex duct installation involves providing adequate support (e.g., strapping maximum 5 ft apart) and ensuring no excessive bends or compression trying to squeeze through narrow spaces or between framing.

It is also important to ensure proper refrigerant charge because it significantly affects the energy efficiency of HVAC equipment. HVAC installers should be accountable for using one of the accepted industry refrigerant charge testing protocols and documenting the results. If weather conditions are too cold to allow for testing, TXV valves should be required. TXV valves located in the outdoor compressor unit automatically adjust refrigerant flow to help compensate for improper charge.

The last part of a quality-installed HVAC system is to provide pressure balancing so air can circulate from bedrooms back to central returns when doors are closed. This has become a significant issue because HVAC systems predominantly use central returns in new homes (Figure 4.13). Without pressure balancing, bedrooms will go into positive pressure when doors are closed as air supply continues to flow without a return path to the central return duct. This can result in

Figure 4.13: Pressure Balancing Problem.

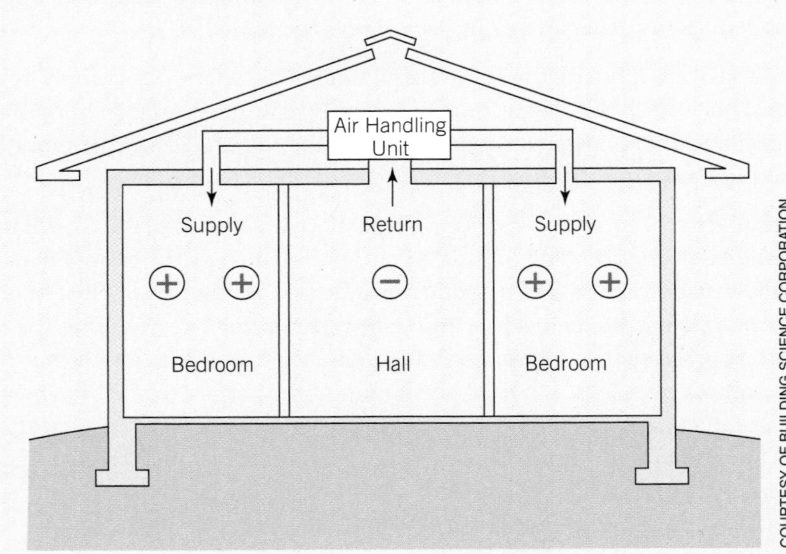

inadequate space conditioning in bedrooms as pressure builds and restricts flow. At the same time, central rooms will go into negative pressure as HVAC fans continue to pull in air at full-rated power without access to bedroom air. The resulting positive pressure in bedrooms can force air flow into remaining small holes in the adjoining construction assemblies, and the resulting negative pressure in central spaces can draw outdoor air into construction assemblies. This uncontrolled air flow can lead to air quality, health, and durability issues.

Typically pressure balancing is achieved using either transfer grills or jump ducts. Transfer grills are simple through-the-wall grills located in bedroom walls adjoining a hall or above the bedroom door. They are sized for adequate air flow from the bedroom to the hall (e.g., 1 sq in. opening for each 1 cfm of air flow) and usually include baffles for acoustic and visual privacy. Jump ducts, as the name implies, are short ducts that jump over the bedroom door through the attic with a ceiling grill on either side (e.g., one in the bedroom and one in the hall). There appears to be a homeowner aesthetic preference for jump duct ceiling grills rather than transfer grills, so jump ducts are much more commonly used by builders who address pressure balancing. However, each jump duct introduces two additional holes into the unconditioned attic that hurt overall performance by allowing additional conduction losses to hot or cold attics. Another option for pressure balancing is the old system of providing dedicated return ducts from each bedroom instead of a central return. This is the most effective option for pressure balancing, but it is least preferred by builders due to the extra costs.

Summary of Building Science

The strategies for controlling air flow, thermal flow, and moisture flow recommended throughout the discussion of building science rules are summarized in Table 4.1. Together they represent a comprehensive package of measures for constructing homes that work.

High-Performance Homes Include Energy-Efficient Components

Once a home is constructed with comprehensive building science measures, the energy loads are so small that other loads for water heating, lighting, appliances, and plug loads become a much more significant part of total energy consumption. As interest grows in net zero homes that use relatively expensive renewable energy systems to offset on-site energy use, energy-efficient components become an increasingly important complement to high-performance homes.

HVAC Equipment. The reduced first cost for smaller right-sized HVAC equipment and smaller, compact duct systems should be applied to high-efficiency heating and cooling equipment. As a minimum, all equipment should be ENERGY STAR qualified.

Water Heating. Efficient water heating options have historically been very limited. This is due to a relatively short efficiency range for the most common gas-fired tank water heaters (.56 to .65 EF) and electric resistance tank water heaters (.88 to .92 EF). Much more efficient integrated gas water heating and space heating systems are available (.90+ EF), but they are very costly. However, instantaneous gas water heaters have become more widely available in recent years and provide much greater efficiency (.80+ EF). Although they are costly, they are attracting

Table 4.1: Comprehensive Building Science

CONTROL AIR FLOW	CONTROL THERMAL FLOW	CONTROL MOISTURE FLOW
Seal unintended holes • Air sealing[a] • Complete air barriers • Duct sealing **Provide intended holes** • Whole-house ventilation • Spot ventilation • Direct-vent combustion **HVAC quality installation** • Duct sealing • Right-sized equipment • Right-sized ducts • Right-sized terminals • Ducts in conditioned space • Properly installed ducts • Proper refrigerant charge • Pressure balancing	**Conduction** • Insulation R-value • Insulation installation (no gaps, voids, compression) • Minimize thermal bridging • Low-e windows **Convection** • Air sealing[a] • Complete air barriers • Insulation alignment • Duct sealing **Radiation** • Low-e windows • Radiant roof barriers[b]	**Moisture vapor** • Air sealing[a] • Complete air barriers • Duct sealing • Vapor barriers or retarders • Whole-house ventilation • Spot ventilation • HVAC quality installation **Bulk moisture**[c] • Heavy roof membranes (eaves and valleys) • Roof flashing details (drip edge, kick-out) • Weather resistant barriers • Pan flashing (windows and doors) • Fabric filter (foundation drain pipe) • Capillary breaks (foundation) • Site drainage

[a] Primary holes are chases, cracks, penetrations, access panels, and drywall at top plates.
[b] Primarily cost effective where HVAC ducts and air handler are located in the attic. However, best practice is to locate all HVAC ducts and air handler in conditioned space.
[c] As a group, these measures are referred to as "water managed construction."

consumer interest because they can provide an endless supply of hot water. On the electric side, progress is being made with heat pump water heaters that approximately double the efficiency of electric resistance water heating. However, they also are very costly and have a history of reliability problems that manufacturers will need to demonstrate are successfully resolved. The most efficient gas and electric water heating options now qualify for ENERGY STAR. In addition, consumers in hot and moderate climates should consider solar water heating. Although solar electric systems appear to have more "sex appeal" and consumer attention, solar domestic hot water heating is a proven technology that is much more cost effective.

Lighting. One of the most cost-effective energy-efficient components to put in high-performance homes is high-efficacy lighting. The primary option is compact fluorescent lamps (CFLs) that fit in conventional fixtures and dedicated fixtures using pin-based bulbs. This lighting uses about one-fourth the energy of incandescent bulbs while reducing wasted heat about 80 percent and lasting 6 to 10 times longer. The waste heat from inefficient lighting can be a significant energy penalty in hot climates and is not beneficial in cold climates where it is much more efficient to produce heat with an energy-efficient furnace, boiler, or heat pump. Concern has been raised about the mercury released in the event a CFL bulb breaks, but many experts suggest the fears are unwarranted because of the trace amounts included in bulbs. The mercury issue may disappear with increased availability of solid state or light emitting diode (LED) lighting. Solid state lighting employs computer chip technology to produce very long-lasting and efficient lighting. The main issues being addressed by the lighting industry are color rendition, optics, and cost, with substantial progress being made on all fronts. As a result, solid state lighting is expected to dominate the market in the next few years.

Appliances and Fans. Other components that complement high-performance homes include energy-efficient refrigerators, dishwashers, clothes washers, ceiling fans, and exhaust fans. All of these products are available with ENERGY STAR certification. In addition, there are high-efficiency options for electric cook-tops and clothes dryers. Electric induction cook-tops transfer heat electromagnetically to ferrous iron cookware with much greater efficiency than electric resistant coil models and offer superior performance (e.g., speed and safety). As mentioned earlier, new heat pump clothes dryers are available in Europe and should find their way to the U.S. market soon. They use much less energy and do not need to be vented outdoors because the moisture is condensed out of the air flow and piped to the same drain used by the clothes washer. There may be issues concerning drying time with heat pump clothes dryers, but they provide direct energy savings from heating much less air and indirect energy savings by eliminating ongoing heat loss and gain caused by the dryer vent to outdoors (see Figure 4.33).

Electronics. Homeowners can equip their homes with much more efficient electronics such as televisions, cable television set-top boxes, phones, and DVD players that also come with the ENERGY STAR label. Primary energy savings come from reducing surprisingly significant energy consumption when these products are not in use yet continue to draw power to keep internal electronic components ready to operate. This consumption has been called "vampire loads" to call attention to hidden energy consumption that few consumers knew about. In fact, one research paper published by Florida Solar Energy Center estimated that the vampire loads for electronics at an entertainment center add up to 220 watts of continuous consumption (5.2 kWh/day).[3] Typically, energy savings are achieved with much more efficient transformers or power modules that keep electronics ready to activate quickly in response to remote controls. New ENERGY STAR specifications for televisions will for the first time also address other improvements that can significantly reduce the large energy load of operating large-screen, high-definition televisions.

---

[3] Danny Parker, David Hoak, Alan Meier, and Richard Brown, "How Much Energy Are We Using? Potential of Residential Energy Demand Feedback Devices," Florida Solar Energy Center, August 2006, http://www.fsec.ucf.edu/en/publications/pdf/FSEC-CR-1665-06.pdf

Table 4.2: Comprehensive Pollutant Control

Source Control	Dilution	Filtration
Moisture • Air sealing • Complete air barriers • Duct sealing • Vapor barriers and retarders • HVAC quality installation • Water managed construction **Chemicals** • Low or no formaldehyde products • Low or no VOC products **Combustion gases** • Direct-vent equipment • CO alarms • Vented fireplaces **Radon** • Radon resistant construction **Biological contaminants** • Rodent screens • Integrated pest management	**Ventilation** • Whole-house ventilation • Spot ventilation (baths and kitchen) • Garage ventilation	**Filters** • Minimum Merv 8 HVAC filter • HEPA filter vacuum or central vacuum

High-Performance Homes Control Pollutants

Following the rules of building science will result in very air-tight and well-insulated homes that are properly ventilated, equipped with effective filters in the HVAC system, and pressure balanced. This is a great start for ensuring a healthy indoor environment. However, pollutants included in building materials and surrounding soils, and introduced by occupants also need to be minimized. Comprehensive pollutant control involves a three-part strategy: source control, dilution, and filtration (Table 4.2).

First and most important is controlling the sources of pollutants, including moisture (e.g., bulk and vapor), chemicals (e.g., formaldehyde and volatile organic compounds [VOCs]), combustion by-products, radon, and biological contaminants (e.g., dust mites, rodents, and bug feces). Moisture is a pollutant because, where excessive, it can substantially increase the risk of mold and dust mites associated with epidemic levels of respiratory disease in American households. In fact, the EPA estimates that at least one person per household has asthma in 19 percent of U.S. homes.[4] Moisture is controlled through building science mea-

[4] "Constructing Improved Homes with Indoor airPLUS," U.S. EPA Indoor airPLUS podcast transcript, December 22, 2010, www.epa.gov/indoorairplus/podcast/constructing_improved _homes.txt

sures that address air flow and from quality installation of HVAC equipment that ensures proper humidity levels. Bulk moisture is controlled with building science measures addressing water managed construction.

VOCs and formaldehyde are minimized by selecting products with no or minimum amounts of these carcinogenic chemicals. Typically these include low-VOC paints, carpets, carpet pads, adhesives, and cabinets. In addition, board products (e.g., plywood, MDF, OSB) should be specified with much safer exterior-grade phenol formaldehyde rather than the more potent urea formaldehyde.

Combustion by-products are controlled by ensuring properly vented appliances but are virtually eliminated by using direct vent equipment and fireplaces with outdoor combustion air (see Building Science Rule No. 10). Carbon monoxide sensors should still be located in all bedroom areas, including homes without combustion equipment, if there is an attached garage.

Radon is a naturally occurring radioactive gas in soils, particularly those with high granitic content. The EPA estimates that 1 in 15 U.S. homes has excessive levels of radon dangerous to occupants.[5] Radon levels can be effectively controlled using radon resistant construction (Figure 4.14). This begins with a plastic sheeting and gas permeable layer under the foundation slab. Both of these are typically used in high-performance homes as part of providing a capillary break for water managed foundations. The plastic sheeting blocks the diffusion of radon through the slab, and the gas permeable layer (4 in. of gravel) allows radon gases to be collected by a plastic vent pipe. This vent pipe starts below the slab with a short "T" fitting to collect any gases, and then typically extends up through a wall in the house, to the attic, and out the roof. A junction box is provided in the attic so a continuous operation fan can be easily installed in the pipe if high radon levels are still evident with the passive venting system. Note that the radon resistant construction system also functions as a subslab depressurization system that continuously removes moisture vapor in the soil.

Biological contaminants are minimized with a diverse set of measures:

- **Air-tight construction using building science practices that limit the penetration of pests**
- **Screens to block pathways for rodents at all vents, except the clothes dryer exhaust vent to avoid trapping lint**
- **Moisture control measures using building science practices that control excessive humidity levels in which dust mites thrive**
- **Integrated pest management (IPM) that provides chemically safe protection from a wide variety of pests and termites**

[5] "A Citizen's Guide to Radon," EPA, December 21, 2010, www.epa.gov/radon/pubs/citguide.html

Figure 4.14: Radon Resistant Construction.

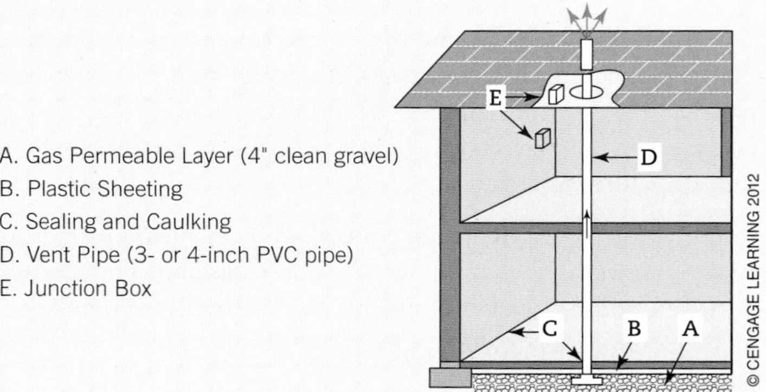

Source: "Buying a New Home: How to Protect Your Family from Radon," U.S. EPA, April 1998, http://www.epa./radon/pubs/rrnc-tri.html

The second part of pollutant control is to dilute a wide array of pollutants occupants introduce into their homes including flame retardants used to treat furniture, heavy metals in home electronics, lots of nasty stuff on the soles of our shoes, candle soot, cigarette smoke, and toxic compounds in air fresheners and cleaning products. In addition, humidity is generated indoors from human respiration, cooking, cleaning, and washing. Dilution is achieved with whole-house and spot ventilation systems in accordance with the American Society of Heating, Refrigeration, and Air Conditioning Engineers (ASHRAE) national standard for whole-house and spot ventilation (ASHRAE 62.2). See Building Science Rule No. 9 for further recommendations on whole-house and spot ventilation. In addition, ventilation fans should be provided in garages to exhaust residual carbon monoxide from car exhaust and dangerous outgases from other stored materials including paints, gasoline, solvents, and lawn chemicals.

The final step in pollution control is filtration. It is typically accomplished by using an effective filter in the return-air duct of the HVAC system. More advanced media and electrostatic filters can filter particulates in the air down to one micron or less in size. In addition, builders should recommend that homeowners use high-quality HEPA filter vacuums or provide a central vacuum system to avoid dispersing dust particles inside the home.

High-Performance Homes Are Disaster Resistant

Homes built using comprehensive building science measures can be expected to have a long life, perhaps hundreds of years. However, unless home construction also addresses locally relevant natural disaster risks, this durability could be undermined. There are many disaster risks in virtually every location, including regionally specific extreme climate conditions (wind, hurricanes, tornadoes, hail, and severe winter weather), natural events (floods, wildfires, and earthquakes), and structure-destroying pests (termites and carpenter ants). Key recommendations for maximizing disaster resistance are included in Table 4.3.

How High-Performance Homes Got Here

Home Performance Was Not an Issue Until the late 1970s

As discussed earlier, homes constructed before insulation started to be installed in the 1950s and 1960s didn't experience significant durability problems associated with moisture damage. The old craftsman builders knew how to keep bulk water from penetrating roofs and walls with weather barriers and flashing details. Basements were another matter with extremely porous old stone foundations. But basements were not commonly finished at this time, and homeowners just took their lumps dealing with occasional standing water from storms and summer humidity. Moreover, moisture flow that got inside construction assemblies was unimpeded from drying with very leaky construction, fewer sheet board products, and the lack of insulated assemblies. These homes appeared to last forever and inspired the phrase, "we don't build them like we used to" in comparisons with modern production housing. However, these homes experienced significant building science failures relative to comfort and excessive energy use. Owners of these old homes accepted being cold in the winter and hot and sweaty in the summer as part of life.

Insulation and Air Sealing in the 1980s Give Birth to Building Science

Heating bills that were once manageable began to increase in the 1980s on the heels of the two oil embargos in the 1970s. This led to insulation and air sealing retrofits in older homes. Their porous foundations still allowed excessive moisture flow, but the homes could not dry nearly as fast. So more comfort and lower bills were exchanged for less durability and greater risk of poor indoor air quality. As a result, many of the old homes that had energy-efficiency upgrades may not last forever after all. This has led to the "first do no harm" mantra cited before improving the performance of existing homes.

In the case of new homes, increasing energy prices influenced energy codes to require more insulation starting in the 1970s. Martin Holladay, a building science expert and reporter, identified five milestones for superinsulated homes that were achieved by 1985:[6]

- **Engineers recalculated the optimum insulation R-values in response to the 1970s energy crises.**
- **Researchers discovered the profound importance of air-tight construction, which led to the development of the blower door for measuring air tightness.**

[6] Martin Holladay, "The History of Superinsulated Houses in North America," Westford Symposium on Building Science, Westford, Massachusetts, August 4, 2010.

Table 4.3: Disaster Resistance Measures

WEATHER	NATURAL EVENTS	PESTS
Wind (all regions) Continuous and adequate load path from the roof to the foundation to transmit uplift and shear loads safely to the ground • Roof-to-wall connectors • Interstory connectors • Foundation anchors • Wall sheathing and shear walls • Reinforced roof and gable ends • Water managed roof • Water managed walls **Hurricanes** • Impact resistant openings • Openings rated for wind design pressure • Reinforced overhangs • Cladding material and fastenings • Out-swing exterior doors **Tornado or hail** • Impact resistant roofing • Openings rated for wind design pressure • Tornado shelter **Severe winter weather (areas > 12" annual snowfall)** • Moisture barrier at roof eave • No heat sources in unconditioned attic • All attic access doors insulated and sealed • No uninsulated recessed lighting • Sealed attic penetrations • Attic access doors insulated, sealed, and gasketed • Proper placement of insulation of water lines	**Floods** • Elevate building 3 ft above the base flood elevation (BFE) • No solid walls below lowest floor in certain Coastal Zones **Wildfires (requirements vary by risk)** • Noncombustible materials –Vents at attic and subfloor –Gutters and downspouts –Driveway street number –Siding –Soffit –Undersides of decks or balconies • Driveway or emergency vehicle access with proper clearances from firewood storage and LP gas • Chimney spark arrestor • Tempered windows and doors • Smoke alarms • Sprinkler system • Vegetation clearance **Earthquakes (for high-risk areas)** • Stable house geometry • Limited building material mass • "Fault zone" restrictions • Reinforced foundations • Floor diaphragm direct load paths, blocking, maximum opening • Adequate shear walls • Roof horizontal opening limit • Secured water heaters • Tempered glass or safety film • Flexible gas line connections with automatic shutoff valve • Secured masonry chimneys • Restrain water heater and HVAC	**Termites (high-risk termite zones)** • Structural materials not edible to termites • Treated wood framing **General pests** • Integrated pest management (IPM)

Source: "FORTIFIED for Safer Living © Builder's Guide," Institute for Business & Home Safety, January 2008.

- **Leading niche builders developed ways to build high-R-value walls, reduce thermal bridging, limit air leakage, and ensure complete air barriers.**
- **Researchers monitored the performance of model superinsulated homes, verifying what worked.**
- **Manufacturers developed and began selling blower doors, heat recovery ventilators, and low-e glazing, all instrumental components to complete building science solutions.**

Thus by 1985 all the ingredients of a home-performance movement were in place including:

- **An understanding of the core principles**
- **Effective dissemination of core principles in books (e.g., *The Superinsulated Home* by J. Ned Nisson), journals (e.g., *Energy Design Update*), and government reports (e.g., "Heat Recovery Ventilation for Housing," published by the National Center for Appropriate Technology).**
- **Key products for construction were developed, including insulation, heat recovery ventilation, and low-e windows.**
- **Diagnostic tools including the blower door and duct blaster were available.**
- **A group of charismatic experts was highly effective at communicating the technical lessons and the business case for change.**

However, not much happened during the following decade. This could be attributed to many factors, but Martin Holladay attributes most of this stagnation to a nearly two-thirds drop in the price of oil and slashed funding under President Reagan for a variety of energy-efficiency programs established by Jimmy Carter.

Builder Programs Engage the Housing Industry in 1995

Beginning in 1995 with ENERGY STAR qualified homes, a wide array of voluntary labeling programs for builders began to emerge. These programs effectively teamed up with established building science experts, Home Energy Rating System (HERS) raters, utilities, state programs, and publications to establish high-performance homes in the housing industry. Some of the most prominent builder programs are discussed in the following sections.

Building Science. A number of government and private-sector voluntary building science programs are available to the housing industry. The most prominent of these is a government label called ENERGY STAR promulgated by the U.S. Environmental Protection Agency (EPA). This label is used for more than 60 different product categories, including new homes. Regardless of the product labeled, the widely recognized blue logo (Figure 4.15) is a symbol of energy efficiency significantly above minimum standard requirements that also ensures cost

Figure 4.15: ENERGY STAR Logo.

Source: U.S. EPA.

effectiveness and overall product performance that meets or exceeds consumer expectations. Nonenergy criteria are critical to the success of the ENERGY STAR label because they help avoid the early failures that can be devastating to new technologies. Consider the profound market preference for gas heating over electric heat pumps based on poor comfort with early heat pump systems promoted by electric utilities in the 1970s and 1980s. Similarly, it has been a very difficult, expensive, and long road getting American consumers to accept high-efficient compact fluorescent lamps (CFLs) after early failures in the 1980s promoting CFLs with poor color rendition, flicker, hum, and premature failures.

Today, the ENERGY STAR logo is recognized by approximately 75 percent of consumers nationwide, and more than 85 percent of consumers in markets with strong state or utility programs. ENERGY STAR for homes was introduced in 1995, and by the end of 2009 there were more than 1 million ENERGY STAR qualified homes with nearly 22 percent market penetration nationwide.[7] As building codes and standard practices continue to improve, ENERGY STAR specifications undergo periodic revisions to ensure above-code performance. In 2012, third-generation specifications will take force that ensure the comprehensive building science measures in Table 4.1 are included in every labeled home (Figure 4.16).

EPA achieves two goals by promoting ENERGY STAR qualified homes. Homeowners reduce their utility bills, and risks associated with climate are reduced with less fossil fuel combustion both at homes and at utility power plants. In 2009 alone, ENERGY STAR qualified homes helped American homeowners save more than $270 million on their utility bills while reducing carbon emissions equivalent to removing 450,000 cars from the road.[8]

Another government high-performance home program is Building America promoted by the U.S. Department of Energy (DOE). This is a research program that funds consortia teams to work directly with America's home builders on leading edge building science solutions substantially more efficient than code. Builders who achieve a very aggressive performance

---

[7] "ENERGY STAR Identity Guidelines (Logo Use Guidelines)," EPA, December 22, 2010, http://www.energystar.gov/index.cfm?c=logos.pt_guidelines

[8] Energy Star for Homes Web site, http://www.energystar.gov/homes

Figure 4.16: ENERGY STAR for Homes, Version 3 Specifications.

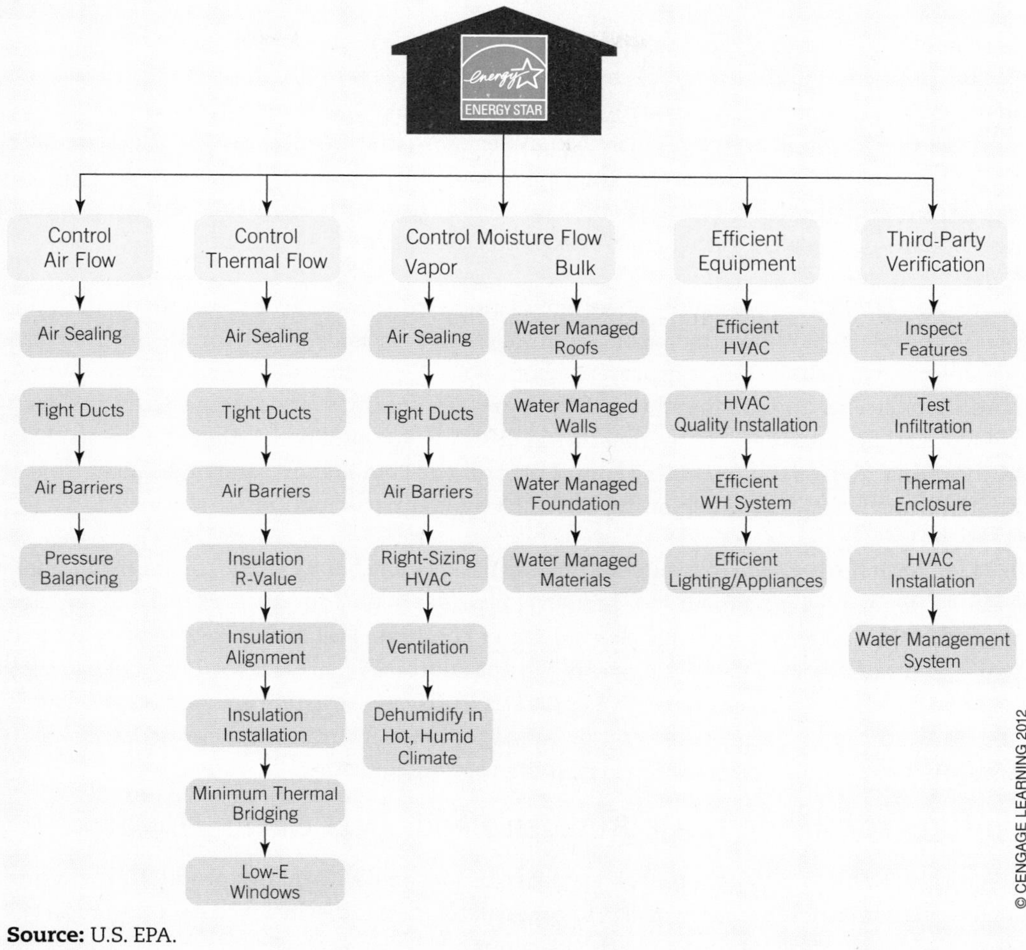

Source: U.S. EPA.

specification are recognized under a subset of this program called "Builder Challenge." Effectively, the Building America program develops, tests, and feeds new building science innovations that complement EPA's ENERGY STAR qualified homes program.

EPA's ENERGY STAR qualified homes and DOE's Builder Challenge are often integrated with private sector building science programs such as Environments for Living by Masco, Energy-Wise by Bayer, and Comfort Home by EIC, Incorporated. These programs offer builders who adopt prescribed building science practices and verification procedures a two-year energy bill warranty for their home buyers. Typically these warranties cover the difference between the actual annual utility bill and a specified amount for each home. The warranted utility bill levels for high-performance homes are typically very low providing participating builders with a marketing advantage.

The Masco and Bayer programs complement their larger parent companies that provide insulation and other building products by encouraging builders to use more of their products, and maybe more importantly, ensuring their products are properly installed. In other words, it appears that manufacturers are increasingly realizing they need to expand their business model to include field application of their products. This is because building trades do not consistently deliver quality installations with established business practices that compensate installers by the piece rather than by the effectiveness of the installation.

Indoor Air Quality. A voluntary government program from EPA, called Indoor airPLUS, ensures every labeled home includes a comprehensive package of indoor air quality measures consistent with the strategy presented in Table 4.2.[9] Almost all of the moisture control, HVAC installation, and combustion safety measures specified are automatically included with a prerequisite that all homes must be ENERGY STAR qualified. Other Indoor airPLUS specifications address the remaining source control and combustion safety measures. This is a relatively new label that is beginning to attract builder participation.

Disaster Resistance. The Institute for Business and Home Safety (IBHS), a nonprofit organization supported by the insurance industry, provides a label for disaster resistance called Fortified Homes. This program addresses disaster conditions relative to severe weather and natural events. Structure destroying pests are not addressed by this label, but I believe they should be included for a comprehensive approach to disaster resistance. Based on this label's affiliation with the insurance industry, discounts for home insurance policies are available to homeowners who purchase homes certified to program guidelines.

High-Performance Home Builder Programs: Green Homes

"Green" has become a widely touted attribute associated with many products, including homes. One commonly accepted goal is to construct homes today in a manner that preserves resources for future generations. At recent count, there were approximately 100 different green labeling programs for homes. The most established national and regional programs include the United States Green Building Council (USGBC) LEED for Homes program, the National Association of Home Builders (NAHB) National Green Building Standard, and the Southface Energy Institute EarthCraft program. All green programs tend to include detailed point schemes, usually with six or seven point categories and multiple-tier thresholds (e.g., bronze, silver, gold, and platinum). This results in complex rating systems with "shades" of green. Thus, the definition of what is "green" varies substantially among all of the programs. In a personal effort to simplify this complexity, I define "green homes" as ones that effectively

9 For information on the EPA Indoor airPLUS program, visit http://www.epa.gov/indoorairplus/

Figure 4.17: Green Home Staircase.

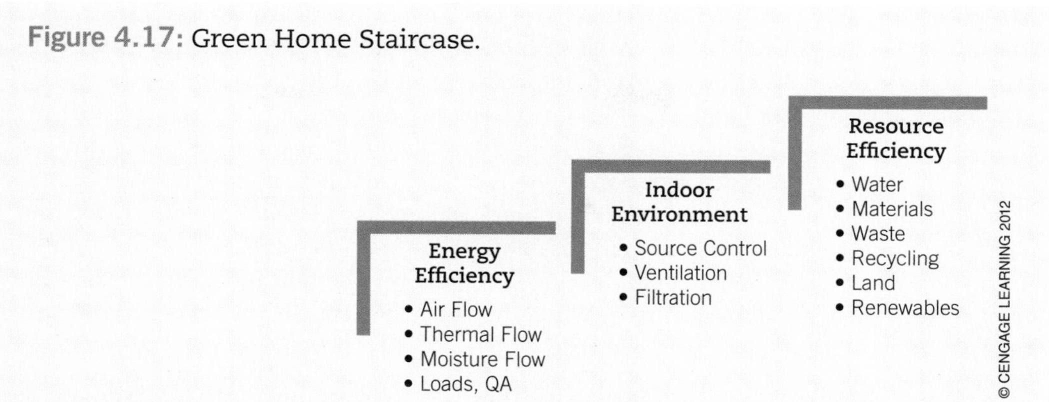

address three major components: energy efficiency, indoor environment, and resource efficiency. Figure 4.17 shows these components along with their key elements as a staircase of priorities.

Energy efficiency is the first and most important component of green because it can significantly reduce the cost of homeownership (e.g., monthly energy savings exceed the incremental monthly mortgage amount attributed to energy-efficient improvements) while providing a critical foundation for improving indoor air quality along with comfort and durability. Almost all green rating systems skew most of the points for energy efficiency because of this cost and performance advantage. Indoor environment should be the next priority because of the obvious importance of protecting occupant health. Finally, resource efficiency addresses the use of land, water, recycling, waste management, sustainable materials, and renewable technologies to most effectively minimize the environmental footprint of new construction. It should be noted that the most critical attributes for green building are also addressed with high-performance homes: energy efficiency, indoor environment, and durability. Thus, high-performance home builders are in a position to easily work within the plethora of green programs or to develop their own proprietary green label.

Why High-Performance Homes Are Broken

Comprehensive Building Science Is Not Applied to New Homes

There is a lack of commitment to comprehensive building science in the housing industry, even though it delivers affordable, comfortable, healthy, and durable homes. This is frustrating to observe knowing the complete package of measures in Table 4.1 can easily be incorporated in every new home at lower ownership cost. Again, the small increase in the monthly mortgage for comprehensive building science improvements is easily offset by the monthly utility savings. I know this point has been raised several times, but it bears repeating. And over time this lower cost of ownership will be further reduced because most mortgages are locked in with a fixed interest rate while utility costs keep increasing. Additionally, homeowners could expect significant cost savings from reduced maintenance expenses (e.g., less moisture damage, less UV damage, longer-lived equipment) and potentially lower health expenses. In addition, there is a huge opportunity cost because many building science improvements (e.g., adding more wall or floor insulation, effectively air sealing, or complete air barriers in wall assemblies) are cost-prohibitive to retrofit once construction is complete. What could be a bigger failure for the housing industry than to ignore a 50- to 100-year opportunity to lock in better performance that costs less to own and results in dramatically less business risk?

A significant part of the reason building science is broken in new homes is that it is not part of the core curriculum in architectural schools and builder training programs. In the absence of building science knowledge and skills, the housing industry has experienced costly failures that are discussed in Chapter 5. However, there is a technology that should accelerate the adoption of building science practices in the housing industry; low-cost infrared cameras. Twenty years ago a typical infrared camera like the one shown in Figure 4.18 cost nearly $30,000, but is available today for under $2,500. As a result, infrared imaging capabilities are becoming much more widely accessible to home inspectors, home energy raters, and energy consultants.

Infrared cameras will have a substantial impact because they effectively expose building science defects to both builders and home buyers in a visually compelling manner. I believe most builders want to build a quality product, but they need clear evidence before making significant changes. Similarly, I believe most home buyers want to purchase a quality-built home, but they need easily understood performance assessments to make informed purchasing decisions. To demonstrate the power of infrared camera technology, the infrared images in Figure 4.19 show the exteriors of two homes in winter. For purposes of a theoretical example, let's assume a home buyer is considering whether to purchase the 1960s home (on left) close to an urban center, or a new high-performance home (on right) in an outlying area.

Infrared images depict surface temperatures of objects. Most commonly, light colors indicate warmer surfaces and dark colors indicate colder surfaces. Thus an exterior image in cold weather with lots of light colors reveals unwanted internal heat loss and possible moisture problems (e.g., condensation where warm interior air flow has reached cold exterior surfaces). Similarly, an interior image in hot weather with lots of light colors reveals unwanted heat gain and possible moisture problems (e.g., condensation where warm, moist air has reached cold drywall surfaces). The bright colors throughout the image of the used home in Figure 4.19 reveal extensive air flow, thermal flow, and moisture flow defects not evident in the new high-performance home. Ergo, a dramatic difference in quality between the used home and the new high-performance home is now obvious to even the least knowledgeable consumer.

Figure 4.18: Low-Cost Infrared Camera.

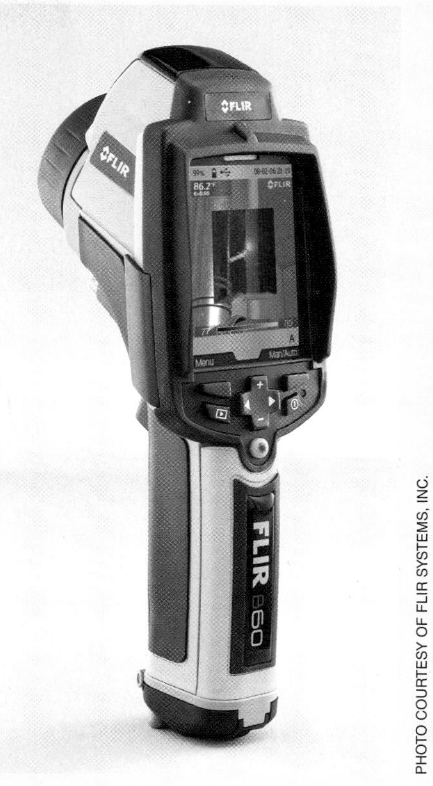

PHOTO COURTESY OF FLIR SYSTEMS, INC.

I want the house on the right, and probably so do you. In fact, I suspect we would all be willing to spend thousands of dollars more or consider a less-preferred location for that visibly superior quality. This peace of mind is especially important because homes are often the largest purchase of a lifetime. With infrared cameras, we can easily recognize the value of a new high-performance home that employs comprehensive building science measures. Builders are typically obsessed with first cost but will come to understand that investments in thermal defect-free homes will yield a compelling value proposition to attract the substantially smaller universe of home buyers to new homes (see Chapter 1). Ultimately, the 100+ million existing single-family homeowners will be forced to consider "deep energy" retrofits to keep their homes from becoming visibly obsolete as shown in infrared diagnostics. *Deep energy retrofit* refers to a major intervention augmenting existing construction assemblies

Figure 4.19: Comparative Infrared Image Exposing Defects.

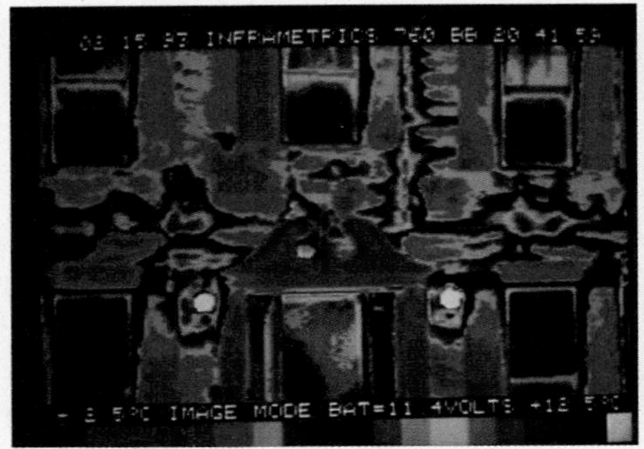

(a) Older home with extensive flow defects

(b) New high-performance home

Source: "Module 1: Getting the Thermal Enclosure System Right," EPA ENERGY STAR Qualified New Homes, November 18, 2009, www.energystar.gov/ia/partners/bldrs_raters/downloads/Thermal_Enclosure_Presentation.wmv.

with comprehensive building science home improvements. In anticipation of these impacts on new and existing housing, I envision that infrared cameras will rapidly become common in the real estate transaction process both for used home inspections and sales of new high-performance homes (see Chapter 6).

Air Flow Is Not Consistently Controlled in New Homes

New homes are too often constructed with excessive air leakage, which increases the risk of high utility bills, comfort problems, reduced indoor air quality, and condensation inside wall assemblies. Some examples follow. Figure 4.20 shows the significant air, thermal, and moisture flow with poorly sealed sill plates and rough openings around doors and windows that still occur in new housing.

Recessed lighting penetrations into attics are another significant source of air leakage due to the large holes and huge driving forces created by superheated air inside fixtures when lights are turned on. Even when insulated can, air-tight (ICAT) recessed fixtures are used, there are often gaps between the sheetrock and the trim that can cause air leakage. This defect is widespread as the housing industry has increasingly used recessed lighting over the past few decades in response to consumer preferences and their low cost.

Another significant source of air leakage that has only recently become evident as a common problem is the gap between sheetrock and top plates at walls adjoining unconditioned attics. This can be attributed to the increased moisture content and less dimensional stability of wood framing than in years past, and increased code requirements for hurricane and

Figure 4.20: Leakage at Cracks.

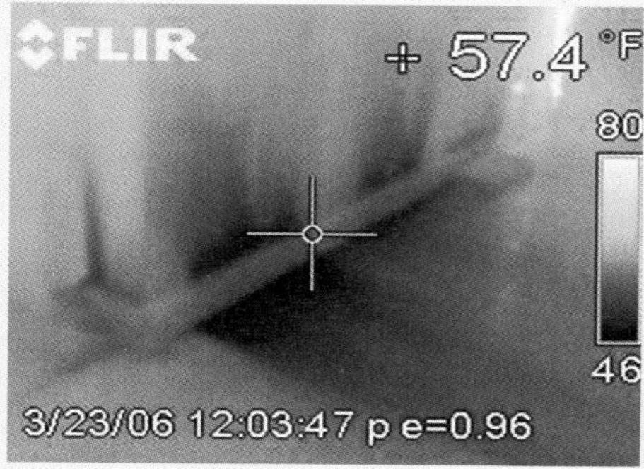

(a) Sill plate leakage

(b) Door opening leakage

© FLIR CORPORATION

Source: "Module 1: Getting the Thermal Enclosure System Right," EPA ENERGY STAR Qualified New Homes, November 18, 2009, www.energystar.gov/ia/partners/bldrs_raters/downloads/Thermal_Enclosure_Presentation.wmv.

tornado straps linking roof framing to exterior walls. The substantial differences in temperature, moisture, and pressure between the attic and conditioned rooms below result in significant driving forces and air leakage across these gaps (Figure 4.21).

Air leakage problems also can be attributed to the "one out equals one in" rule with air pressures created by the many fans now used in homes:

- **Clothes dryer fans rated typically between 150 and 250 cfm**
- **Kitchen cook-top exhaust fans rated from 100 to 2,500+ cfm**

Figure 4.21: Leakage at Drywall Connection to Top Plate.

© CENGAGE LEARNING 2012

Source: "Module 1: Getting the Thermal Enclosure System Right," EPA ENERGY STAR Qualified New Homes, November 18, 2009, www.energystar.gov/ia/partners/bldrs_raters/downloads/Thermal_Enclosure_Presentation.wmv.

- **Whole-house fans rated at 2,500 to 6,000 cfm**
- **Central vacuum fans rated at approximately 100 cfm**
- **Bathroom exhaust fans rated at 70 to 150 cfm**
- **HVAC fans typically sized at 400 cfm/ton (a 1,600 cfm fan for a 4-ton heat pump can create significant house pressures when supply or return duct leakage occurs or pressure balancing solutions are not provided)**

The air flow control challenges associated with the clothes dryer fan are one good example of how problems develop. Air is drawn into the dryer cabinet where it is heated, blown over wet clothes, absorbs moisture, and is then exhausted outdoors through a duct. This cycle continues until the clothes are dry or the machine drying cycle time ends. A clothes dryer with a 200 cfm fan operating for a 60-minute drying cycle will exhaust 12,000 cubic feet of air to the outdoors (200 cfm × 60 minutes). One out equals one in means that 12,000 cubic feet of air must be pulled in from outdoors. This is approximately all the air in a 1,500 sq ft home. Where will this air come from? The easy answer is the path of least resistance. What that path will be varies for each home.

If the clothes dryer is located in a laundry room adjoining a garage where the access door weather stripping is missing or poorly maintained and sill plates not fully sealed, a large portion of this makeup air could be drawn in directly from the garage. Unfortunately, garage air can be a source of many pollutants, including residual car exhaust, out-gassing fumes from stored liquids (e.g., solvents, fuels, and paints), dust, and pests. For example, 50 ppm of carbon monoxide have been measured in garages hours after a car has been started and pulled out when the EPA considers only 9 ppm safe.

If the clothes dryer is next to a gravity exhaust water heater, the 4 in. exhaust flue represents a very big "path of least resistance" hole (Figure 4.22). Add a leaky HVAC air handler cabinet and return duct in the room and the resulting negative pressure can potentially reverse the natural draft exhaust of carbon monoxide back into the room. In some cases, the negative pressure is large enough to induce flame roll-out at the bottom of the water heater, as is evident by the burn stains at the bottom of the hot water tank in Figure 4.22. This can be a serious combustion safety problem.

In Figure 4.23 a family room wall adjoins a laundry room, and the negative pressure induced by the clothes dryer fan flow is drawing hot attic air into the wall at the air gap between the drywall and the top plate. The resulting hot surface temperatures at this wall can be expected to cause discomfort and high utility bills.

These potentially significant home performance problems are associated with just one of the many fans now included in homes, the clothes dryer fan. Consider the potential problems presented by the many other fans included in new homes. Air-tight construction is not

Figure 4.22: Clothes Dryer Induced Air Flow at Water Heater.

Figure 4.23: Clothes Dryer Induced Air Flow at Leaky Top Plate.

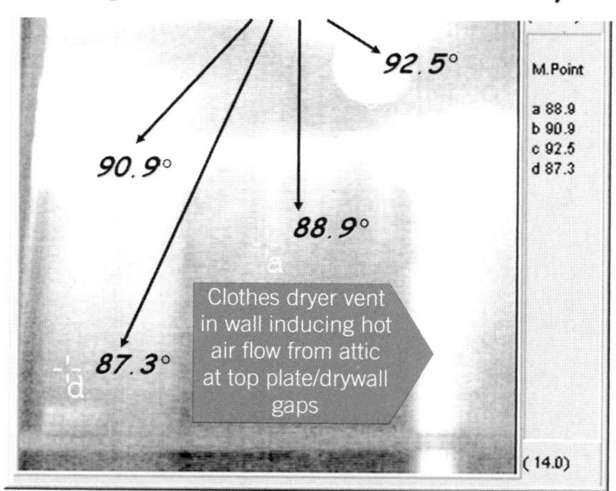

consistently achieved in new home construction, increasing the risk of combustion safety, poor indoor air quality, comfort problems, increased moisture, and high utility bills.

Insulation System Is Not Properly Installed in New Homes

Although building codes have significantly increased the quantity of insulation since the 1970s, it could be argued that the requirements are still inadequate. This is because it is cost-prohibitive to upgrade construction assemblies with a complete insulation system. Thus, every new home should include following measures with minimal defects to effectively control thermal flow (see Building Science Rule No. 5):

- **Adequate insulation at or above the national energy code**
- **Proper installation without gaps, voids, compressions, or misalignments**
- **Complete air barriers to block thermal bypasses**
- **Minimal thermal bridging to avoid excessive conductive losses**

Unfortunately, infrared imaging diagnostics reveal an epidemic of thermal defects in the nation's existing housing stock. To a large extent, this could be attributed to the use of lowest cost insulation materials that require greater quality control while typical business practices compensate installers by the piece rather than performance. This has been especially egregious with batt insulation, which experiences the most problems with gaps, voids, compression, and misalignment. This includes failure to split batt insulation around wiring and plumbing, excessive compression at walls and band joists, and complete misalignment at ceilings above unconditioned spaces (e.g., garages and cantilevers). It is easy to find examples of poorly installed insulation such as those shown in Figure 4.8. Under pressure to meet very competitive price targets, the housing industry has swept this issue under the rug. Infrared cameras will have a dramatic impact by exposing common defects with standard installation practices so that they can no longer be ignored.

Another common problem is missing or ineffective insulation and gaskets at attic access openings such as ceiling hatches, drop-down stairs, and knee-wall doors. This results in significant thermal flow at these openings (Figure 4.24).

An example of systemic misalignment in housing construction is the common practice called inset stapling. This involves fastening the tabs of paper-faced batt insulation to the inside face of wall studs. Significant air spaces between the drywall and insulation and outside sheathing and insulation can result as highlighted in gray in Figure 4.25. This misalignment can lead to significant convective heat transfer as follows. The air space adjoining the interior wall will assume near room temperature due to little thermal resistance from drywall, and the air space adjoining the exterior wall will assume near outdoor temperature due to little thermal resistance from exterior sheathing and siding. Based on Building Science Rule No. 1, there will be a substantial driving force from more to less heat in hot and cold weather

Figure 4.24: Example of Poor Insulation at Attic Access.

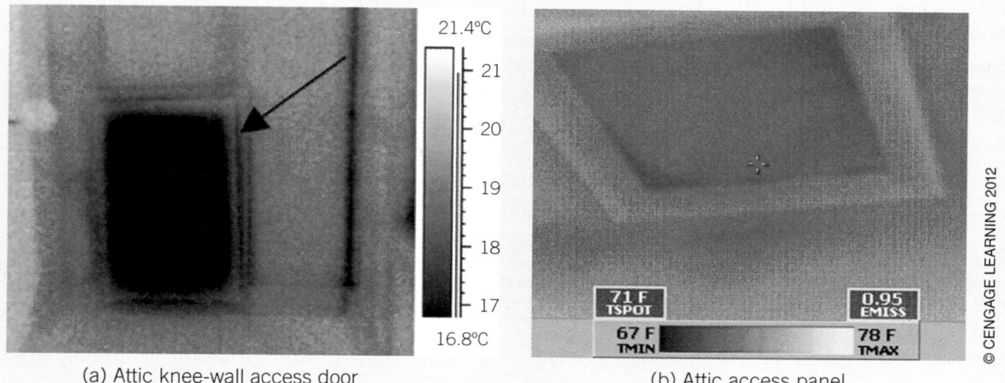

(a) Attic knee-wall access door

(b) Attic access panel

Source: "Module 1: Getting the Thermal Enclosure System Right," EPA ENERGY STAR Qualified New Homes, November 18, 2009, www.energystar.gov/ia/partners/bldrs_raters/downloads/Thermal_Enclosure_Presentation.wmv.

Figure 4.25: Convective Loop Heat Transfer with Inset Stapling.

Source: "Module 1: Getting the Thermal Enclosure System Right," EPA ENERGY STAR Qualified New Homes, November 18, 2009, www.energystar.gov/ia/partners/bldrs_raters/downloads/Thermal_Enclosure_Presentation.wmv.

resulting in a convective loop that bypasses the thermal resistance of the insulation (see arrows in Figure 4.25). Ostensibly inset stapling is done to accommodate drywall contractors who want to glue and nail drywall to the exposed face of studs. However, drywall can easily be installed with the insulation tabs fastened to the front face of the studs (face stapling). This allows the insulation to be in complete contact with the drywall and outside sheathing without compression for substantially better performance.

Once again an infrared image provides compelling evidence of the building science failure. Figure 4.26 shows the outside of a home with inset stapled insulation on a cold winter day. The image reveals a substantial thermal bypass due to the convective loop. Small voids that appear bright white at the top and bottom of the wall function like air passages that dramatically accelerate thermal flow around the insulation. This is a good example of how small defects can have large impacts on insulation effectiveness.

Another component of a complete insulation system that is substantially broken in new home construction is complete air barriers. In particular, the industry does not consistently address the need to provide a six-sided air barrier assembly for all fibrous insulation applications. When fibrous insulation is installed without a complete air barrier assembly, discoloration will often reveal the thermal bypass failure. As an example, Figure 4.27 shows discolored insulation at an attic knee wall due to the missing backside air barrier. This discoloration is visual evidence of significant thermal bypass with the insulation functioning like a high-efficiency filter as air flows through it.

Figure 4.26: Winter Infrared Image of Home with Inset Stapling.

Source: "Module 1: Getting the Thermal Enclosure System Right," EPA ENERGY STAR Qualified New Homes, November 18, 2009, www.energystar.gov/ia/partners/bldrs_raters/downloads/Thermal_Enclosure_Presentation.wmv.

Figure 4.27: Missing Backside Air Barrier at Attic Knee Wall.

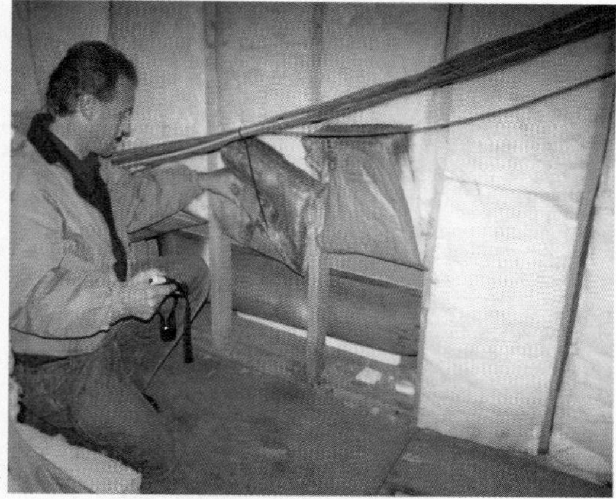

COURTESY OF BLUE GRASS ENERGY

Conceptual diagrams showing thermal bypass at a dropped ceiling are great for explaining building science principles, but infrared images are much more effective when demonstrating the problem. Figure 4.28 shows an infrared image of a new home with extensive dropped ceilings that are missing a lid below the attic insulation. The resulting thermal flow control

Figure 4.28: Thermal Bypass Problem at Dropped Ceilings.

Source: "Module 1: Getting the Thermal Enclosure System Right," EPA ENERGY STAR Qualified New Homes, November 18, 2009, www.energystar.gov/ia/partners/bldrs_raters/downloads/Thermal_Enclosure_Presentation.wmv.

Figure 4.29: Air Barrier Missing at Duct Chase.

Source: "Module 1: Getting the Thermal Enclosure System Right," EPA ENERGY STAR Qualified New Homes, November 18, 2009, www.energystar.gov/ia/partners/bldrs_raters/downloads/Thermal_Enclosure_Presentation.wmv.

failure is revealed by the pervasive cold surface temperatures at all assemblies adjoining the dropped ceilings. This will dramatically compromise comfort, lead to high energy bills, and increase the risk of moisture problems.

Another critical air barrier detail is the cap at all chases. The infrared image in Figure 4.29 shows one example in which there is excessive air flow from the hot attic into a home due to an incomplete air barrier on top of a duct chase. Note that this thermal bypass is enhanced by a leaky return duct that creates negative pressure or suction when the HVAC fan is operating.

A final example in which air barriers are commonly missing is band joists. Figure 4.30 shows a typical installation in which batts are crammed into the small framing spaces. In addition to the obvious gaps, voids, and compression, there is no air barrier facing the interstitial space (air space between the finished ceiling and subfloor above). This is particularly problematic in cold climates where the room temperature air in this space has many clear pathways to the cold outside sheathing that can lead to condensation.

Figure 4.30: Band Joist Insulation with No Air Barrier.

Thermal bridging is also a significant problem in new construction. Although many production builders are moving to panelized construction, which helps minimize framing, much of the industry still uses an excessive number of studs, headers, and plates. Figure 4.31 shows an example of the almost unconscionable extremes sometimes observed. On top of significant compromises to thermal performance, excessive framing adds cost for materials, labor, and waste removal.

Finally, a word about thermal flow control with radiant barriers. During a trip to Bermuda, it was great to observe the ubiquitous white roofs that impress tourists. Somehow, the inhabitants have figured out that it is smart and nearly free to substantially decrease cooling loads with reflective roofing. However, in hot climates in the United States, minimal attention is paid to reflective roofs. A preference for colors other than white is no excuse because many reflective roofing materials are now available in a variety of colors. At least there is an increased use of radiant barrier faced roof sheathing in some hot climate regions. The housing industry is also embracing control of radiant heat flow with low-e window technology thanks to building codes and programs like ENERGY STAR.

Figure 4.31: Example of Construction with High-Framing Factor.

HVAC Systems Are Poorly Installed in New Homes

There is virtually no infrastructure for installing HVAC systems to minimum standards by manufacturers and industry associations. Instead, HVAC systems commonly are not properly sized, installed by unskilled tradespeople equipped with cheap ineffective filters, and not fully tested and diagnosed at completion for proper air flow and refrigerant charge. As a result, systems consistently fail to deliver full rated performance and equipment life. One study evaluating recent research projects on HVAC system installation practices found on average air conditioning systems are oversized in 47 percent of installations, air flow is inadequate in 70 percent of installations, refrigerant charge is improper (under- or overcharged) in 74 percent of installations, and duct leakage is excessive averaging 270 cfm at 25 Pascal test pressure or nearly 20 percent of fan flow for an average 3.5 ton system with a 400 cfm per ton air handler. This study estimates that HVAC systems correcting these poor installation practices could be expected to save 24 to 35 percent of energy use.[10]

One of the reasons HVAC trades are reluctant to properly size equipment is that they have been held accountable for building science problems not associated with their work. For example, one of the most common call-back problems in the housing industry is the bonus

[10] Chris Neme, John Proctor, and Steve Nadel, "National Energy Savings Potential from Addressing Residential HVAC Installation Problems," 1999, pp. 4, 7, 8, 10, and 17, http://www.socalgas.com/calenergy/docs/hvac/references/proctornationalstudy/pdf

room over a garage that is too cold in winter and too warm in summer. Typically the HVAC contractor is called in to fix the homeowner complaint when this problem can most often be attributed to improperly installed insulation and missing air barriers (e.g., floor insulation not aligned with subfloor and no air barrier on back side of attic knee walls). This leaves us with a housing industry that mistakenly thinks "bigger is better" when it comes to HVAC equipment. Thus it is common practice today for equipment to be sized based on some combination of simplified calculations, guesswork, and rough rules of thumb.

Duct sizing is almost completely ignored and often left for unqualified installers to figure out in the field. And sizing for terminals (e.g., supply grills) is not even on the housing industry's radar screen. Yet high-performance homes with extremely low space conditioning loads require proper terminal sizing to effectively distribute their very low air flow in each room.

Beyond sizing, a critical problem with HVAC installations is the common practice discussed earlier of locating ducts and equipment in unconditioned spaces (e.g., attics, crawl spaces, and garages). Attics are the industry-preferred location for HVAC systems in hot climates and increasingly in cold climates as well. This practice compromises affordability, comfort, health, and durability, all in the name of lowest first cost.

Problems with ducts in unconditioned space are compounded by excessive duct leakage. Most HVAC installers have gotten the message that duct tape is completely ineffective, and it is fortunately being abandoned. However, too many contractors still rely on UL 181 tape, which is better but much less preferred than sealing with mastic including fibrous tape at more rigorous connections. Return duct leaks in unconditioned spaces can draw in pollutants, and supply duct leaks in unconditioned spaces waste conditioned air and can lead to condensation at cold surfaces. During the winter, supply duct leakage in attics also can contribute to ice damming where warm air leakage melts snow on the roof that then freezes and forms icicles when it drains down to the cold eaves. This increases utility bills while compromising comfort, air quality, and durability. Figure 4.32 shows an example of poor duct sealing details that need to be eliminated from the industry. This leakage is a problem even when ducts are located inside the conditioned space because it can compromise room-by-room comfort and potentially allow warm air flow to reach cold surfaces (e.g., air leakage in interstitial spaces).

In addition to ducts being leaky, the quality of their installation is extremely poor in most parts of the country. This is especially the case with the most common system used today, flex ducts. They are made of an inner and outer plastic lining around a metal coil for shape and to hold insulation. In other words, we use something with the robustness of a plastic garbage bag as the comfort delivery system in homes worth hundreds of thousands of dollars. Flex duct typically comes in a standard 25 ft length with a variety of diameters and insulation levels. Their corrugated interior profile inherently restricts air flow, which gets things off to a bad start. Common installation defects make it much worse. First, excessive duct lengths are often used because installers avoid cutting them to minimize assembly work. This results in numerous unnecessary bends that further restrict air flow. Widespread

Figure 4.32: Example of Leaky Ducts in New Construction.

installation defects compound the problem, including hard bends greater than the diameter of the duct (e.g., hard turns at boot connections), compression (e.g., squeezing ducts through narrow spaces at truss joists or between framing members), and insufficient strapping, which creates sagging and additional bends.

Typical HVAC systems are also incomplete because they are missing whole-house ventilation and pressure balancing. As a result, homes are not equipped with an adequate fresh air supply system and can experience excessive positive and negative pressure conditions that increase the risk of comfort and durability problems.

All of these HVAC system installation problems are not just the HVAC subcontractors' fault. Home builders have been beating down vendor prices for years, and they in turn have had to compromise on training and investments in quality practices. Recognizing how broken HVAC system installations are across the housing industry, ENERGY STAR for Homes is adding HVAC quality installation as a new requirement with Version 3 specifications that take full effect on January 1, 2012. It will be a heavy lift to ramp up the entire HVAC installation infrastructure across the country.

Water Management Is Not Comprehensive in New Homes

Could you imagine the response from any prospective homeowner if a builder were to ask, "Where would you like me to leave the risk of moisture damage in your new home?" Of course the response would be, "I don't want any risk of moisture damage!" But home buyers are not given that choice, and the housing industry continues to leave out critical parts of a complete water management system. Which parts are left out vary by region and builder, but too often parts are missing. Some of the more prominent missing measures can include one or more of the following details:

- **Kick-out flashing where roofs meet walls.** They cost less than $10 per detail and can avoid thousands of dollars in damage from water draining behind siding.
- **Heavy roof membranes at valleys.** There can be a hundred times more concentration of water than at a point higher up the roof.
- **Heavy roof membranes at eaves.** They provide icing protection in climates that have freezing weather.
- **Weather resistant barriers.** They need to be provided in all new homes, and properly installed with proper shingle fashion lapping, cap nails instead of staples, and repairs to cuts and tears to ensure full drainage behind cladding.
- **Pan flashing at windows and doors.** This is not an established practice even though 100 percent of windows and doors leak.
- **Deck framing fully flashed to the exterior wall weather resistant barrier.** Anchor bolt connections are a significant point of entry for moisture that need to be fully flashed.
- **Drain tile wrapped in fabric filter.** In their absence, foundation drains can clog with sediment after seven or eight years.
- **Capillary breaks under slabs and foundation stem walls.** In their absence, 12 pounds of moisture can diffuse through foundations each day.

The foundation details (drain tile wrapped in fabric filter and capillary break) are the most egregious omissions because they are virtually impossible to retrofit after construction is complete and can compromise water management for a home that should last hundreds of years. Additionally, the capillary break under the slab serves double duty as the most important component of a radon resistant construction system that should never be omitted in EPA Radon Zones 1 and 2.

Homeowners dread any risk associated with wet basements, musty smells, mold, and moisture damage and would therefore value the added protection from comprehensive tried-and-true water management practices. The industry is obsessed by first cost and does not think they can sell these features (see Chapter 6).

Use of Energy-Efficient Components Needs to Be Increased

The housing industry can do a much better job integrating high-efficiency components in new homes. Gravity exhaust combustion equipment is still too frequently used, especially when located in conditioned spaces that require outdoor combustion air. The combustion

Figure 4.33: Wall with Dryer Vent Hot from Attic Thermal Flow.

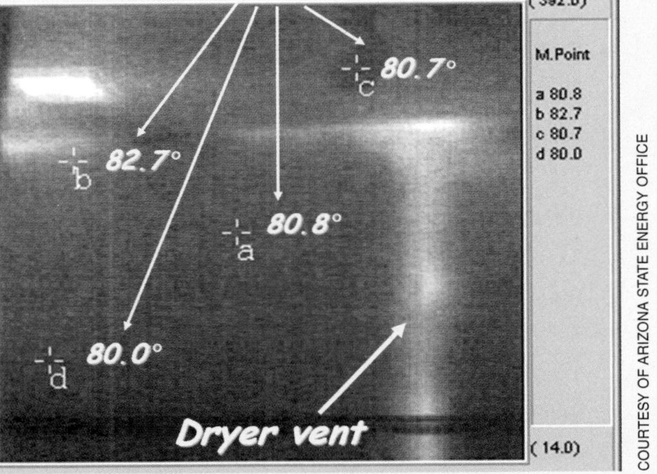

COURTESY OF ARIZONA STATE ENERGY OFFICE

safety and energy-efficient benefits of direct-vent equipment should never be traded off for lower first cost. Additionally, the housing industry is deficient in equipping homes with high-efficacy lighting (bulbs and fixtures). The significant energy savings, excellent performance, and reduced maintenance should not be ignored. The housing industry is getting better at offering ENERGY STAR appliances in homes, but they still need to be more routinely provided as standard equipment. And the industry needs to do a better job looking to upcoming technologies. For instance, the heat pump clothes dryer, even though not a typical builder-supplied appliance, can effectively address the thermal defect caused by the vent inside walls and ceilings. To demonstrate this defect, Figure 4.33 shows an infrared image of a family room wall adjoining a laundry room, taken during the summer. The image reveals that the dryer vent in the wall is a significant source of "more to less" heat flow from the attic when the dryer is not operating. In the winter, the wall area around the dryer vent would show a cold surface temperature. The heat pump clothes dryer, currently popular in Europe, avoids this defect because it is ventless.

Pollutants Are Not Effectively Controlled in New Homes

The most significant broken practice for ensuring good indoor air quality in new homes is the minimal effort made to control sources of pollutants. This starts with incomplete building science practices addressing moisture flow (both bulk and vapor) to minimize the risk of mold and dust mites. Additionally, the industry fails to consistently specify products that minimize dangerous chemicals, including board products, cabinets, paints, carpets, and adhesives without formaldehydes and volatile organic compounds. As mentioned earlier, the risk of back-drafted combustion by-products from space heating and water heating equipment can be completely eliminated with direct-vent options that are too often not used to minimize first cost. In addition, the risk of radon can be effectively addressed with simple, low-cost radon resistant construction practices commonly missing, even in regions rated for high exposure.

As mentioned under the section on HVAC systems, whole-house ventilation is commonly missing in new homes. Without effective fresh air delivery systems, pollutants are not being consistently diluted. In addition, garages are rarely equipped with exhaust ventilation to dilute residual car exhaust and potentially dangerous fumes from stored gasoline, paints, and solvents.

Lastly it is common practice to equip HVAC systems with cheap filters that do a poor job eliminating particulates from the airstream. Although much better filters are widely available; they are not routinely considered in HVAC system design and specifications.

Disaster Resistance Is Not Addressed in New Homes

A point that bears repeating is that the recommendations in this book for retooling the housing industry will result in homes that should last hundreds of years. It would be foolish not to invest in added disaster resistance so these homes can benefit from this substantially enhanced durability. However, the housing industry has not embraced comprehensive disaster resistance practices. The lack of attention is becoming increasingly evident in the aftermath of severe hurricanes, tornadoes, hail storms, severe winter weather, floods, fires, and earthquakes all across the country. Much of the excessive damage can be attributed to not specifying impact resistant materials, not elevating homes adequately above base flood elevations, not using appropriate noncombustible materials and clearances in fire regions, and not constructing with termite resistant structural systems. Moreover, homes are exposed to greater disaster risks as development keeps spreading out to more marginal land on flood plains, adjoining wildfire-prone areas, in close proximity to earthquake faults, and in regions with significant termite populations.

How to Fix High-Performance Homes

Integrating high performance in new homes is not easy. There are many concepts, details, and value propositions associated with building science, energy-efficient components, pollutant control, and disaster resistance. The following recommendations recognize that it is cheap insurance for the housing industry to make major investments in all of these critical components.

Commit to High-Performance Homes from the Top Down

Home performance is not extra credit; it is mission critical. Plain and simple, any builder that is not building high-performance homes will soon become irrelevant. The housing industry should anticipate a quick transition to rational consumer behavior with a profound market preference for high-performance homes that cost less to own because they uniquely address the critical home buyer concerns:

- **Fear of rapidly increasing energy costs**
- **Desire for quality construction associated with thermal and acoustic comfort**
- **Low maintenance with better moisture control, disaster resistance, and higher-grade components**
- **Elimination of dangerous chemicals in daily life**
- **Assurance of a good investment with advanced technology construction that delivers superior performance not possible with older homes**

This market preference will be further reinforced when the financing and insurance industries finally catch up to the fact that high-performance homes significantly reduce risk and respond with discounted mortgage and insurance rates. "Used homes" will become functionally obsolete. Minimum-code new homes won't be able to compete.

This is a value proposition the housing industry needs to embrace to compete in the "new normal" with dramatically fewer home buyers and location preference shifting to urban centers (see Chapter 1). The housing industry has made critical mistakes by ignoring home performance, and this has led to costly failures (see Chapter 5).

While it is time to make a full business commitment to home performance, this can only be accomplished from the top down. The chief executive with every builder needs to gather every decision maker in the company and install a sense of urgency and accountability with all aspects of constructing high-performance homes. This includes initial planning, design, construction, and sales. Once home buyers experience the superior quality, there will be no turning back. Along the way, builders will realize substantial cost savings as they climb the learning curve, experience increased demand for advanced new construction, significantly lower customer service center costs, and dramatically lower risk. But it takes the top person to say this is who we are as a company.

Invest in Risk Reduction

I love sleeping at night knowing I've minimized any potential failures in my work. The housing industry needs to adopt the same philosophy by investing in quality details that eliminate the risk of mold, moisture, and customer dissatisfaction. The prevailing lowest first-cost mentality driving the industry will have to be sacrificed for investments in assured performance. Long-term profits should be effectively maximized because it costs less to fix mistakes before they happen. Installing exterior insulation sheathing in cold climates is a good example. Yes it costs more, but it typically moves the dew point outside of the exterior sheathing, eliminating the possibility of moisture condensation problems inside the construction assembly. I will pay for that risk reduction. Consider the much greater customer satisfaction from lower bills, improved comfort, and superior noise reduction a nice bonus. Similarly, I will pay for furring to create an air gap between the cladding and the sheathing. With this air gap, walls will positively drain, mitigating the risk of bulk moisture problems. It is true that home buyers cannot see these and many other details recommended in this chapter for high-performance homes, but there are many ways builders can create experiences around these quality improvements that will resonate with their buyers (see Chapter 6). I don't want the cheapest construction and neither do home buyers. It's time to ignore silly statistics about how many more families cannot afford to buy a home with each incremental dollar of first cost. This is especially important because you can pay for all these improvements with the monthly cash-flow from lower utility bills, along with cost savings from right-sized homes (see Chapter 2) and lean construction practices (see Chapter 5). The culture of the housing industry needs to change from lowest first cost to maximum performance and quality construction.

Invest in Comprehensive Building Science

There is no shortcut; all of the measures shown in Table 4.1 should be included in every retooled home. However, there is an easy solution: construct homes conforming to ENERGY STAR Qualified Homes Version 3 specifications (see Figure 4.16). This will ensure a complete thermal enclosure system, HVAC quality installation system, and water managed construction system.

Basic thermal enclosure requirements for insulation at or above code, air-tight construction, and low-e windows are making inroads with the housing industry. However, new commitments will have to be made to address widespread problems with improperly installed insulation and reduced thermal bridging. It's time to deliver thermal defect-free homes.

The housing industry needs to make a special effort to adopt HVAC quality installation practices. This is a complete paradigm shift for the existing HVAC installation infrastructure. This begins with locating all equipment and ducts inside the conditioned space. Then quality installation needs to hold installers accountable for all aspects of the HVAC system, including proper sizing of equipment, ducts, and terminals; ensuring matched indoor and outdoor components; installing effective distribution systems field tested for proper air flow and duct tightness; verifying proper refrigerant change; including a whole-house fresh air ventilation system; and equipping systems with filters that effectively remove particulates. Finally, builders should provide spot ventilation and pressure balancing. It is time to deliver engineered comfort with fresh air and effective filtration in every home.

Comprehensive water managed construction is needed to completely drain water from the roof, walls, openings, site, and foundation. This is critical with comprehensive building science because the construction assemblies cannot dry effectively after getting wet. Most of these measures will not be visible, but home buyers will quickly appreciate the quality and peace-of-mind advantages if they are effectively integrated in the sales process (see Chapter 6). It's time to provide better water protection in new construction.

However, this is all just the beginning. The housing industry also needs to quickly plan to exceed ENERGY STAR specifications with even greater air tightness, above code levels of insulation, super windows, and balanced ventilation. The age of net zero homes is fast approaching, and the industry needs to be prepared by first including all 100-year opportunity cost measures that cannot be cost-effectively added after construction is complete.

Own the Holes

Immediately after joining ENERGY STAR for Homes, I began to travel extensively to meet with builders to benchmark housing industry performance across the country. One trip in January 1995 took me to Chicago during intensely cold and miserable winter weather. In the middle of a blizzard, I drove a small compact car, with a woefully undersized heater, into the night to meet a special builder. Even though it had been an extremely long day and I was frozen and barely knew how to find my way back to the hotel, I was not going to miss this visit. The builder, Perry Bigelow, was so special because all of his home buyers received an energy bill warranty that their townhomes would cost no more than $200 to heat for the entire winter season. Since he started this warranty a number of years ago at $100 per heating season, he had paid out only a few claims attributed to historically cold winters. When I visited Perry Bigelow, I asked him how he could make such an extraordinary warranty. His answer was that he "owned the holes."

If the name of the game in real estate is "location, location, location," the name of the game in building science, especially for cold climates, is "holes, holes, holes." Effectively sealing the holes can negate lots of other evils in home performance. Perry knew this and had only one trade person directly employed by his company, the air sealing specialist. His job was to own all the holes, which entails effectively air sealing all penetrations, cracks, and chases, including installing air barriers at all thermal bypass conditions. This is unusual for an industry that typically outsources all work to the various trades. That is a mistake when it comes to sealing holes in high-performance homes. First, it's difficult to get consistent quality work on this critical component to good building science. Second, in-house capabilities are needed to correct work by a multitude of other trades who receive no compensation for sealing holes or maintaining air-tight construction. For example, plumbers and electricians get no additional compensation for neatly cutting and sealing holes for piping and wiring.

The reason air leakage and complete air barriers are so important is that both are essential to all three critical elements of building science: air flow, thermal flow, and moisture flow. Air sealing is the one job that should be the responsibility of the builder. It's time to hire an in-house skilled craftsman whose compensation is directly linked to air-tight performance on all projects, measured and verified with blower door testing and infrared camera diagnostics.

Invest in Energy-Efficient Components

It's time to make energy efficiency a primary criterion for all products specified in new homes. Typically they provide higher-quality construction, better performance, and longer warranty protection in addition to their energy savings. Energy-efficient components that should be part of every high-performance home include HVAC equipment, water heating, appliances, fans, and lighting. Builders should also consider including clothes washers and dryers with new homes when heat pump dryers become more widely available as an opportunity to eliminate the dryer vent. Finally, builders should educate buyers about equipping their high-performance homes with other energy-efficient products, including a wide array of consumer electronics.

Invest in Pollutant Control

With complete building science, homes are two-thirds of the way to comprehensive pollutant control: water managed construction; air-tight construction; tight ducts; right-sized equipment, ducts, and terminals; whole-house and spot ventilation; pressure balancing; and direct-vent equipment. The housing industry should use EPA's Indoor airPLUS label to take them the rest of the way (similar to the list of measures in Table 4.2). This small extra investment will allow builders to market a comprehensive indoor air package: continuous fresh-air system; effective filtration of particulates; moisture control that prevents mold and dust mites; no dangerous chemicals in carpets, paints, cabinets, framing, and adhesives; no combustion by-products from heating equipment; and radon protection. I believe this is an untapped value proposition because the same U.S. population that spends over $10 billion per year on bottled water[11] and over $20 billion per year to keep dangerous chemicals out of their food[12] should be willing to spend pennies per day to ensure healthier air for their families. Assuming indoor air quality measures add less than $2,500 to the cost of an average home, the incremental increase in the monthly mortgage is about $12, or 40 cents per day. It's time for the housing industry to offer American home buyers a healthier indoor environment for pennies a day. It's also time to find out that home buyers will gladly pay for it.

Invest in Disaster Resistance

The housing industry needs to embrace prudent disaster protection measures. Homes on or near flood plains should be raised at least 3 ft above the base flood elevation with no solid walls below the first floor. Homes in hurricane-prone areas should consider integrating working shutters that can quickly and easily be closed to protect windows during storms as opposed to relying on nailing plywood over openings. Homes in heavy wind locations should have reinforced roof construction, heavy-duty connectors from roof to

[11] "Report: Bottled Water Increasingly Comes from the Tap," *USA Today*, August 12, 2008, http://content.usatoday.com/communities/greenhouse/post/2010/08/bottled-water-tap-supplies-/1

[12] Carolyn Dimitri and Lydia Oberholtzer, "Marketing U.S. Organic Foods, Trends from Farms to Consumers," U.S. Department of Agriculture, Economic Information Bulletin No. 59, September 2009, http://www.ers.usda.gov/publications/eib58/

foundation, and heavy gauge roofing materials. Homes in tornado zones should have safe rooms and exterior finishes with greater impact resistance. Homes in locations with severe winter weather should have no heat sources in the attics and fully vented roofs that always stay cold to avoid ice dams. Homes in areas exposed to wildfires should be constructed with noncombustible materials and appropriate clearances. Homes exposed to earthquakes should use stable house geometry, limited high-mass building materials, reinforced shear walls, flexible gas line connections, and secured masonry chimneys. And homes in high-risk termite locations, especially southern Gulf Coast states, should be constructed with termite-resistant framing and materials. The easiest way to ensure comprehensive disaster resistance (see Table 4.3) is to work with the IBHS Fortified Homes program. Homes constructed to survive disasters will speak volumes about the core values of the new retooled housing industry.

Table 4.4: High-Performance Homes

What It Is	How It Got Here	Why It's Broken	How To Fix It
High performance has four components that ensure affordability, comfort, good indoor air quality, and durability: 1. Building science • Air flow • Thermal flow • Moisture flow 2. Efficient components • HVAC equipment • Water heating • Lighting • Appliances • Fans • Electronics 3. Pollutant control • Source control • Dilution • Filtration 4. Disaster resistance • Weather • Natural events • Pests	Home performance was not an issue until the late 1970s. Insulation and air sealing in the 1980s gave birth to building science. Builder programs engaged the housing industry starting in 1995: • Building Science: – EPA ENERGY STAR for Homes – DOE Building America – Manufacturer warranty programs • Indoor Air Quality: – EPA Indoor airPLUS • Disaster Resistance: – IBHS Fortified Home • "Green" Homes: – USGBC LEED for Homes – NAHB National Green Building Standard – Southface EarthCraft program	Comprehensive building science is not applied to new homes. Air flow is not consistently controlled in new homes. The insulation system is not properly installed in new homes. HVAC systems are poorly installed in new homes. Water management is incomplete in new homes. New homes need more efficient components. Pollutants are not effectively controlled in new homes. Disaster resistance is not addressed in new homes.	Commit to high-performance homes from the top down. Invest in risk reduction. Invest in comprehensive building science. Own the holes. Invest in energy-efficient components. Invest in pollutant control. Invest in disaster resistance.

Chapter 4 Review

So What's the Story?

It is extremely expensive and virtually impossible to retrofit homes for comprehensive high performance once construction is complete. This is why high-performance homes will radically change the value proposition for the housing industry. Once again it will become a compelling choice to buy a new home, but not for short-lived cosmetic improvements. Instead, high-performance homes will provide life-changing improvements in affordability, comfort, health, and durability. Moreover, these improvements will pay for themselves with energy savings and deliver impressive peace of mind. The high-performance home story can be summarized as follows:

- **What It Is.** High-performance homes apply technologies and practices that consistently ensure affordability, comfort, good indoor air quality, and durability.
- **How It Got Here.** High-performance homes began when insulation and air sealing practices emerged after two oil embargoes in the 1970s and matured with a wide array of builder programs after 1995.
- **Why It's Broken.** Substantial evidence reveals widespread performance problems with the nation's housing stock: thermal defects, air leakage problems, dangerous levels of pollutants, and poor disaster resistance.
- **How To Fix It.** Top-down commitments are needed to make substantial investments in comprehensive building science, energy-efficient components, pollutant control, and disaster resistance.

The details are included in Table 4.4.

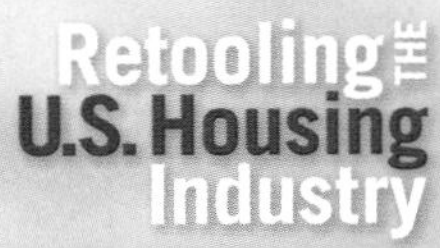

5

Quality Home Construction: Stop Protecting Old Technologies and Practices

QUALITY HOME CONSTRUCTION: PROCESS, GOALS, AND HOW GOALS ARE ACHIEVED

Process. *Construction details and practices vary significantly, particularly from small custom builders to large production builders. However, the process generally entails the following steps:*

- ***Prepare Construction Documents.*** *These documents are prepared after the design process is completed for a custom or model home. They effectively transfer the intent of the final design to detailed plans that meet the minimum requirements of the local building department. Generally, this includes floor plans, elevations, sections, details, and necessary specifications.*
- ***Secure Permits.*** *Upon completion, the construction documents are submitted to a local building department for a building permit. Environmental, school, utility connection, and other requirements vary by locations and also must be met.*
- ***Negotiate and Sign Contracts.*** *Material lists (e.g., concrete, framing, windows, doors, siding, roofing sheetrock, lighting, cabinets, finishes) are generally prepared by the builder, designer, or purchasing manager. The construction documents and material lists are used to negotiate purchase agreements with suppliers and subcontractors.*

Quality Home Construction

Goals
- On budget
- Defect-free
- Efficient processes

How
- Construction docs
- Techs./systems
- Quality assurance
- Lean production

© iStockphoto.com /miflippo

© iStockphoto.com /jhorrocks

- ***Construct Home.*** When permits are secured, the construction schedule begins with site and foundation work and ends with finishes and landscaping. A series of scheduled inspections by the building department must be passed for the project to progress. When construction is complete and verified in compliance with all building code requirements, a certificate of occupancy is issued by the building department.

Goals. *For most builders, first cost appears to be the number one goal as they strive to complete construction at or below an established budget for maximum profit. This budget is based on an arranged price for presold homes or an expected sales price for "spec" homes. The next goal is to deliver quality construction consistent with the specified scope of work. Specific concerns include workmanship, material selection, and home performance. These concerns are directly linked to customer satisfaction and reduced callback repair work. A further goal is to employ efficient processes to ensure construction comes in on budget, on time, and with minimum waste.*

How. *Builders need a comprehensive process for employing best technologies, practices, and systems. This begins with construction documents that include critical details and specifications needed to contractually enforce consistent quality work. Best available technologies and practices that support building science and indoor air quality objectives (see Chapter 4) should be employed. It is common to get in a comfort zone protecting old technologies and practices, but builders should fully evaluate and apply innovations that can solve real problems and minimize risk. The housing industry needs to employ effective quality assurance procedures that ensure all scopes of work are consistently met as efficiently as possible. Finally, lean construction methods should be used to minimize construction costs associated with time, level of effort, safety risks, quantity of materials, waste, and jobsite trips.*

This chapter reviews past technologies that have had the most significant impact on the housing industry, then identifies underutilized technologies, practices, and processes that now need to be embraced by the housing industry to achieve the goals for quality home construction.

What Is Quality Home Construction?

A number of different processes address quality home construction. Building codes do a reasonably good job of ensuring structural integrity, leak-free plumbing and gas lines, safe electrical systems, and fireproof assemblies. Other components of construction under the oversight of building codes (e.g., energy efficiency, mechanical systems, and finishes) are subject to varying degrees of enforcement. Quality construction requires a comprehensive business strategy for making sure the best available technologies and practices are employed, work is performed with minimal defects, and waste is minimized at every step in the process. A builder who just meets code requirements has provided the worst home allowed by law. The goal is to ensure quality far beyond the minimum level of oversight provided by local building departments, including conforming to design specifications and high-performance home standards at the lowest cost. This goal is achieved with two key elements: quality assurance and lean production. Although often assumed to be similar concepts, they differ much the way "management" differs from "leadership." Quality assurance, like management, gets things done right; and lean production, like leadership, gets the right things done.

Ensuring the Right Technologies and Practices Are Employed

The design process addresses quality issues associated with selecting preferred grades of trim, finishes, hardware, cabinets, and lighting and plumbing fixtures (see Chapter 3). Integration of high-performance materials, equipment, and installation practices are specified to ensure comprehensive thermal enclosure systems, quality installed HVAC systems, water management systems, pollutant control, and disaster resistance (see Chapter 4). The construction process needs to systematically include the most cost-effective technologies that can minimize risk and maximize performance. Typically this is achieved with details, specifications, and notes included in the construction documents. This provides a legal framework for holding all work accountable to the prescribed scopes of work.

Ensuring Work Is Done Right

Quality assurance applies a comprehensive system for effectively managing the complex home construction process. The six-step process presented in Table 5.1 is based on personal experience and participation in housing research projects. This discipline is relatively new to the housing industry, and there are many variations on the theme. However, many of the key principles presented are consistent with recommendations by different experts, often with origins in total quality management (TQM) concepts. TQM was developed by W. Edwards Deming, who worked with Japanese industrialists after World War II. One of the most important principles states: it is much easier and less costly to ensure quality work is built into the process than to relegate it to inspection upon completion.

Table 5.1: Six-Step Quality Assurance Process

Quality Assurance Step	Objective
1. Selection of Materials and Practices Fully consider best available technologies and innovations.	Ensure building science principles are fully addressed, product ingredients are safe, and there is a clear track record or sound science for durability.
2. Documentation Include specification notes and details in contract documents and other agreements with subcontractors.	Provide detailed scopes of work that clearly specify standards for materials and workmanship.
3. Checklists Provide critical specifications for each major work category that clearly spell out all requirements and standards of conformance.	Provide a tool that accompanies the installation process to facilitate consistent implementation of all prescribed measures.
4. Training Disseminate skills for installation, testing, inspection, and selling specified materials and practices that conform to quality standards.	Prepare all workers and sales agents so they are fully prepared for success when installing materials to quality standards and when selling the advantages of quality-built homes.
5. Testing and Inspections Complete tests and inspection by third-party or in-house staff to measure conformance of installed work to specified standards.	Enforce accountability for all work in conformance to prescribed standards and with minimal defects.
6. Root Cause Analysis Use diagnostic processes to evaluate the origins of any defects uncovered during testing and inspections and identify solutions that address the cause.	Instill a formal process for correcting any uncovered defects so they are permanently eliminated from the construction process.

Ensuring Minimum Waste and Continual Improvement

Lean production is a continual improvement system that seeks to minimize wasted materials, time, and cost in the construction process. In the early 1950s, lean production principles were developed and successfully implemented by Toyota Motor Company in Japan. Lean construction principles were introduced to the construction industry in 1992.[1] An abridged definition from the Lean Construction Institute (LCI) is that "lean construction is a production management-based philosophy emphasizing the need to simultaneously design a facility and its production process while minimizing waste and maximizing value to owners throughout the project phases."[2] This effectively represents a whole paradigm shift in the housing industry's business philosophy. The current business model determines price by assessing the production cost and then adding profit to it. In contrast, lean production builders determine the price the consumer is willing to pay independent of construction cost, and then calculate profit by subtracting the cost of production from the selling price. Profits are maximized by reducing the cost of production in lean systems. Table 5.2 includes the seven forms of waste[3] identified from the original work at Toyota that remain as critical guidelines today for lean production.[4]

How Home Construction Got Here

Research at the National Association of Home Builders (NAHB) Research Center indicates it takes 10 to 25 years for the housing industry to adopt innovations.[5] This is too long to wait for technologies that can effectively address the great challenges facing the industry today (see Chapter 1). These technologies can dramatically improve quality, reduce cost of ownership, improve comfort, reduce occupant health risks, and increase durability. In addition, they will have profound impacts on reducing builder risk and increasing customer satisfaction. The following sections review major technology and construction practice innovations from their ancient origins through the birth of our nation and on to the present day. A consistent set of categories is used throughout.

U.S. Home Construction: Ancient Origins

The origins of many technologies and practices used in housing today date back to ancient civilizations.

---

[1] Lauri Koskela, "Application of the New Production Philosophy to Construction," Technical Report # 72, Center for Integrated Facility Engineering, Department of Civil Engineering, Stanford University, CA, 1992.

[2] Tariq Abdelhamid, "Lean Production Paradigms in the Housing Industry," Pathnet.org

[3] R. Mastroianni and T. Abdelhyamid, "The Challenge: The Impetus for Change to Lean Product Delivery," Proceedings of the 11th Annual Conference for Lean Construction, Blacksburg, VA, July 22–24, 2003.

[4] Ibid.,Tariq Abdelhamid

[5] C. Theodore Koebel, Maria Papadakis, Ed Hudson, and Marilyn Cavell, "The Diffusion of Innovation in the Residential Building Industry," U.S. Department of Housing and Urban Development; prepared by Center for Housing Research, Virginia Polytechnic Institute and State University and NAHB Research Center, January, 2004.

Table 5.2: Seven Forms of Waste Addressed with Lean Production

FORM OF WASTE	EXAMPLE
1. Overproduction Producing beyond the customer requirements, producing unnecessary materials/products.	Producing more pipe spools than required.
2. Inventory Holding or purchasing unnecessary raw supplies, work-in-progress inventory, finishing goods.	Stockpiling too much drywall in an area well before it is needed and in the way of other trades.
3. Transportation Multiple handling, delay in material handling, unnecessary handling.	Locating materials too far from the point of installation.
4. Waiting Time delays, idle time.	Crew B waiting for an activity to be completed as promised by Crew A.
5. Motion Actions of people or equipment that do not add value to the product.	Double and triple handling of material when planning could have reduced it to one move.
6. Overprocessing Unnecessary processing steps or work elements.	Rubbing a concrete foundation wall too well when it will be backfilled or covered.
7. Correction Producing a part that is scrapped or requires rework or additional procedures.	Punch list items or items of work that are deficient and require rework.

Foundations. Concrete, the dominant material used for foundations today, was first developed more than 5,000 years ago when Egyptians used an early form to build the pyramids. In 100 BC, Romans used concrete remarkably close to modern cement to build many of their architectural masterpieces, including the Coliseum and the Pantheon.

Framing. Buildings in China were supported by wooden frames seven millennia ago. In fact, the Chinese were the first to develop load-bearing timber frame construction, a network of interlocking wooden supports forming the skeleton of the building. This is considered China's major contribution to worldwide architectural technology.[6]

Materials. Stucco and plaster mixtures were developed by the ancient Greeks and Romans. In addition, masonry and stone were widely utilized.

Insulation. The discovery of asbestos for insulation is credited to the ancient Greeks, who named it, but asbestos is critically harmful to human respiratory systems. The Greeks used cavity wall construction to contain insulation, knowing the air trapped between inner and outer wall sheathing would help reduce warm air flow out during cold winter weather and hot air flow in during the summer. The Vikings and other northern Europeans learned to insulate their homes with mud chinking plastered in the cracks between the logs or hewn boards of building walls. Mud mixed with horse or cattle dung and straw was known as daub and was considered a stronger, better material than plain mud. They also used sheep's wool for insulation both for clothing and as a drapery to line the interior walls of their homes. This helped to block drafts from leaky construction, provided a buffer from the low wall temperatures, and soaked up dampness. Similar benefits were derived when embroidered woven tapestries hung on interior walls came to be widely used as insulation in stone castles during the Middle Ages.[7]

Windows. The first primitive homes had doorways to enter and leave but were often built without windows. The advent of the smoke hole to vent smoke from fires used for cooking and heating provided the first "accidental" unglazed opening, allowing daylighting to brighten the dark interior. The first transparent window glass was used in ancient Roman times. The largest known piece of Roman glass was 3 ft by 4 ft, installed in a public bath in Pompeii. The first cast plate glass was not developed until the 1600s in France.[8]

Space Conditioning. Heating in the earliest ancient structures relied on indoor fires, which led to the aforementioned smoke holes for venting. Ancient Egyptians utilized thick brick walls to keep their desert homes and buildings cool. This was an ancient application of thermal mass for purposes of buffering the interior from the intense summer heat (see Chapter 3). This natural comfort architecture was perfected by the Pueblo Indian settlements in the Four Corners region of the American southwest (intersection of Colorado, Utah, New Mexico, and Arizona). In particular, the famous Cliff Palace 200-room village constructed between AD 1100 and 1300 in Mesa Verde, Colorado, employs thick masonry wall construction for homes carved into steep stone cliffs facing south. The cliffs effectively shade the

---

6 Wikipedia, "Ancient Chinese Wooden Architecture," last modified October 7, 2010, http://en.wikipedia.org/wiki/Ancient_Chinese_wooden_architecture

7 Roberto Bell, "A Brief History of Insulation—Look How Far We've Come," Article Dashboard.com, December 7, 2010, http://www.articledashboard.com/Article/A-Brief-History-of-Insulation---Look-How-Far-We-ve-Come/834447

8 John Carmody, Stephen Selkowitz, Dariush Arasteh, and Lias Heschong, *Residential Windows*, 2nd Ed. (New York: W.W. Norton & Company, 2000).

Figure 5.1: Taos Pueblo.

dwellings from the summer sun while allowing the low winter sun to penetrate for passive solar heating. Southwestern architecture is still profoundly influenced by similar high-mass architecture used in the Taos Pueblo constructed by the Tiwa Indians. Constructed between AD 1000 and 1450 near Taos, New Mexico, stepped block houses up to five stories high used thick adobe walls to buffer interiors from the hot summer sun (Figure 5.1).

Plumbing. More than 2,800 years ago, the fabled King Minos of Crete owned the world's first flushing water closet, complete with a wooden seat. Lost for centuries in the rubble of the palace ruins, the invention did not rematerialize until 1594. Two were built by Sir John Harington: one for his godmother, Queen Elizabeth, and one for himself. Unfortunately, another 200 years would go by before the indoor water closet was taken seriously.[9] The ancient Romans were famous for using gravity aqueduct systems to provide water for indoor plumbing and baths.

Electricity. There was no electricity for refrigeration, heating, lighting, or cooking in ancient buildings. As a result, space cooling and heating often relied on proper solar orientation and shading techniques, thermal mass to store solar heat, and cross-ventilation strategies to capture prevailing breezes (see Chapter 3). Heating, cooking, and lighting relied on burning wood inside homes, which indirectly led to the first opening for daylight to vent smoke, the smoke hole.

---

[9] Theplumber.com, "History of Plumbing in America," July 1987, http://www.theplumber.com/usa.html

U.S. Home Construction 1750–1850: Birth of a Nation

At the country's origins, housing was often a very basic architectural box with a steep pitched roof to shed snow and rain. Many early homes had exposed framing on the interior because there was no plumbing, electricity, or insulation that would benefit from cavity wall construction. These homes proved to be very durable as all wood framing could easily dry when it got wet. However, the homes were exceedingly leaky, and uncomfortable during extreme weather conditions.

Foundations. Thick stone or brick walls were used as foundations. These walls allowed water in the soil to seep through, which would drain through the dirt floor. Water management problems are common today in historic homes where stone foundations have been upgraded with slab floors and finished basements.

Framing. Lumber was cut and milled by hand into timber and boards and constructed with timber framing, also known as post and beam construction. In this framing system, timber framing was assembled with interlocking joinery secured with wooden pegs. The exterior sheathing was wood planks finished with wood shingles and shakes.

Materials. Wood and brick were the most widely available materials to finish the exterior of homes. Timber framing was most commonly used, and the interiors were usually exposed lumber and planks.

Insulation. Homes were not insulated. You survived cold winter weather by sleeping under thick covers and gathering around fires. In the summer, natural ventilation was used to enhance cooling.

Windows. Windows relied on cast plate glass, which was difficult to manufacture in large pieces. Thus window mullions were not for decoration but to allow for easier assembly of smaller pieces of glass. Innovations in the production of glass were finally seen in the 1800s, allowing for larger, stronger, and higher-quality glass at a price that made it more available to the general public.

Space Conditioning. For the first 100 years, homes in America were constructed in heavily forested areas. Heating was dominated by burning wood in brick fireplaces and in derivatives of the cast iron Franklin stove, invented in 1742. It was not until 1885 that the nation would burn more coal than wood. The heat produced was usually only for central spaces. The lack of heat throughout the home delayed the development of bathrooms.

Plumbing. Most colonial bathing consisted of occasional dips in ponds or streams. There was no indoor plumbing, and water for cooking and bathing relied on a hand pump and pail. For bathing, water was heated in a pot over a wood fire or in a kettle over the cooking stove. Some stoves had a reservoir lined with tin, copper, or porcelain that could be filled with water. Heating enough water for a bath was so time consuming that people didn't bathe much and often masked body odor with perfumes and oils or just went around odor challenged.

Much of Saturday was spent getting cleaned up for church on Sunday. Outhouses were used for latrines. However, open wells had easy access to contamination from nearby privies, which contributed to increased rates of sickness and death. Early in the 1800s, the stack was vented through the roof, but proper sizing was poorly understood and usually understated. This led to small vent pipes that often clogged up with frost during the winter.

Some cities began to develop waterworks projects to provide water for domestic uses and for firefighting (chimney fires and wood-framed buildings were a bad combination). Early systems relied on gravity fed water starting from a spring or stream on high ground. Prior to steam power in the 1800s, water wheels harnessed energy from river flow to raise the water. The first systems were crude pipelines put together with wooden pipes made of bored-out logs. In 1804 Philadelphia became the first city in the world to adopt cast iron pipe for its water mains. A glimpse of the future appeared in 1829 when the Tremont Hotel in Boston became the first building with indoor plumbing. The hotel included eight water closets and cold running water in bathrooms, the kitchen, and the laundry. However, early water closets would prove to be a source of contamination and a health hazard.

Electricity. Many household appliances we take for granted today were absent due to the lack of electricity. Lighting was provided by candles and oil lamps, and food that would decay could not be stored.

U.S. Home Construction 1850–1920: The Industrial Revolution

Homes remained uncomfortable during this time due to a lack of insulation and drafty construction, but important amenities were added. This included improved central heating, running water, indoor water closets, and electric lights that were found in the homes of the wealthy at the turn of the century.

Foundations. The invention of Portland cement is most often credited to Joseph Aspdin in England in 1824. However, as a construction material, concrete was not ready for prime time until much later when a Parisian gardener, Joseph Monier, invented reinforced concrete in 1867 and Henri Le Chatelier addressed a slow curing rate problem in 1887 by optimizing the chemical reactions of oxide rates. Concrete was still in its infancy during this period, and stone and brick remained the primary foundation option.

Framing. Construction shifted from post and beam to framed cavity walls using a technique called balloon framing. No one is sure who introduced balloon framing in the United States, but the first building to use it was probably a warehouse constructed in 1832 in Chicago. This system uses wood studs to frame walls that extend continuously from the sill plate to the eave line with intermediate floor structures supported on ledger blocks attached to the studs. This requires long length wood studs, which are much more difficult to manufacture, and creates multifloor air flow pathways that exacerbate fire resistance. Exterior sheathing still relied on wood planks.

Materials. The exterior of homes continued to be finished with wood and masonry. However, framed cavity walls were typically finished with plaster.

Insulation. The only insulation in housing constructed during this time period was provided by the air space created with cavity construction. Thus homes continued to be excessively leaky and uncomfortable. As the Industrial Revolution became firmly entrenched, manufacturers turned to asbestos for their insulation needs, much like the ancient Greeks. The primary application was steam-powered equipment with lots of hot pipes. The pipes needed to be wrapped in asbestos to make them safe for workers. With the invention of the steam locomotive, the demand for asbestos exploded. Suddenly fireboxes, boilers, pipes, boxcars, and breaks were lined or wrapped in the dangerous heat retarding, flame-resistant fibers. Asbestos was not used for residential applications until much later.

Windows. The size and number of windows in homes increased dramatically with production innovations, but there was essentially only one type of glass available: clear, single-pane, cast plate glass.

Space Conditioning. By the end of the nineteenth century, two major heating options were available for American homes: boilers and furnaces. Low-cost cast iron radiators were responsible for bringing central heating to large numbers of homes. Coal-fired boilers in basements delivered hot water or steam to these radiators in every room. In 1885 Dave Lennox built and marketed the industry's first riveted-steel coal furnace. Without widespread availability of electricity and fans to move air, these early furnaces transported heated air by natural convection (warm heated air rising) through ducts from the basement furnace to the rooms above. For those homes that did have electricity, Schulyer Wheeler invented the electric fan in 1886. The electric fan became the primary tool for cooling in homes until the post-World War II economic boom.

Plumbing. By 1845, earlier investments installing sanitary sewers began to pay off, providing an outlet for waste water, indoor plumbing, and working water closets, which were starting to make inroads in new housing. By the mid-1850s, finer new homes were being designed with dedicated bathrooms. Unfortunately, bad plumbing and the stench from open sewer connections made some new homes uninhabitable.[10] However, there was a major breakthrough in venting when an unknown plumber suggested balancing the air pressure in the system with the outside atmospheric pressure to prevent siphoning or blowout of the water seal in traps. In addition, plumbers finally learned to increase the size of the vent pipe.

Water supply systems were also advancing. New York completed an aqueduct-fed system in 1842 that transported water from a huge reservoir 40 miles north of the city to secondary reservoirs in the city, including the current Central Park location, and then to a network of underground mains. In 1869 Chicago unveiled a major breakthrough with a twin-tunnel

[10] Theplumber.com, "History of Plumbing in America," July 1987, http://www.theplumber.com/usa.html

system that drew water from 2 miles out in Lake Michigan with coal-fired steam engines. This system provided 15 million gallons per day into the city's water mains. By the 1860s, all but 4 of the 16 largest U.S. cities had municipal water supplies.

Modern sewage systems also had their origin during this period. Engineer Julius W. Adams developed guidelines for a sewer he was commissioned to design for the city of Brooklyn. These principles were published in textbooks and became available for towns and cities across the country. This period brought together the critical components for good plumbing, including proper venting, waterworks, and sewers.

The time for indoor water closets and plumbing had also arrived, although early models continued to be a source of contamination and a health hazard. This includes the American plunger closet introduced in 1857, which was unsanitary. After several iterations of improvements, Robert Frame and Charles Neff of Newport, Rhode Island, produced the prototype of the siphonic washdown closet in 1900. However, there were still failures to develop the necessary flushing action, and this resulted in overflow problems. By 1910, a redesigned bowl by Fred Adee would spur the production of the siphonic closet in America.

The growth of plumbing was extraordinary in the early to mid-1900s as American households were being equipped with water closets, sinks, faucets, and tubs, including many innovations in materials and finishes. However, until this time water still had to be heated on the cooking stove.[11] Then three water heating technologies were introduced following the turn of the century: automatic storage, automatic instantaneous, and solar water heaters.

Starting in the 1920s, a system called range boilers provided automatic storage. These systems had a chamber or pipe loop called a water-back installed in the firebox of the stove. Heated water moved by convection through this chamber to a storage tank. Variations of the range boiler ensued that would eliminate the need to fire up the kitchen stove when hot water was needed, saving fuel and overheating during warm weather.

The first automatic instantaneous water heaters were also introduced at the turn of the century for heating bath water. Water flowing through a sprinkler would spray out through the combustion gases and collect the heat, and unfortunately combustion by-products as well. It would then flow over metal that was being heated by a flame, collecting more heat, and finally run to a spigot and into the tub. One manufacturer boasted in a 1906 *Sweet's Catalog* advertisement that their units were 92 percent efficient.

The third system, solar water heating, was also starting to catch on around the turn of the century. The batch heater, also known as an integral collector storage (ICS) unit, was a simple type of solar system where one or more tanks were placed in a box behind a single-pane glass cover. Water was directly heated by the sun with no moving parts and had inherent

[11] Larry Weingarten and Suzanne Weingarten, "The History of Domestic Water Heating," *PM Engineer* (first appeared in *Home Power Magazine*, August/September 1995), August 29, 2000, http://pmengineer.com/Article/Feature_Article/2000/08/29/The-History-of-Domestic-Water-Heating

freeze protection due to the large thermal mass of the stored water. The main drawback was substantial overnight heat loss, especially in cold climates. Thermosyphon solar water heaters were introduced to address this constraint. Water in tubes behind the glass collector would heat and rise by convection to a tank mounted above the collector.

These water heating on demand technologies introduced in the early 1900s to American housing brought a luxury not even available to royalty until this time. Among these systems, instantaneous water heaters and solar water heating eventually lost market share and disappeared. Instantaneous water heaters had a number of problems, including precise temperature control. Solar water heaters had early leakage problems and could not compete with increasingly available gas at attractive prices. It is only in the last decade that the water heating industry has again diversified with all three types of systems.

Electricity. In 1882 the first coal-fired electric power plant opened in New York City, delivering enough power to light 11,000 lightbulbs. This marked the beginning of the end for gas and kerosene lamps. In 1892 Thomas Edison was awarded a patent for what he called the "electric conductor." Essentially it was the first electric wiring system including insulating wire that was designed to be waterproof and fireproof. Edison believed this technology would make his electric lamp a mainstream household feature. The very earliest wiring systems introduced shortly after his invention are known as "knob-and-tube." Individual conductors were run spaced apart, but the wires were susceptible to dampness and abrasion as they passed through walls and floors. For protection in these places, "insulating tubes" made of porcelain were used. The earliest form of overcurrent protection introduced at this time was the fuse block or fuse holder with a safety plug for easy replacement.

U.S. Home Construction 1920–1950: The Electric Grid Emerges

An increasing number of homeowners would live better than royalty of years past during this time. There was hot water on demand, and space heating was made substantially more convenient with gas- and oil-fired forced air furnaces that used electric powered fans to distribute heated air through ducts to all rooms. Sheet products emerged with the advent of sheetrock, although this material was not widely accepted until the end of World War II.

Foundations. Poured concrete became widely available and was the foundation footing of choice. Foundation walls were transitioning from stone and brick to concrete block and finally to poured concrete.

Framing. Balloon framing was popular when long lumber was plentiful, but this was no longer the case in the early 1900s. By the 1930s, balloon framing was largely replaced by platform framing that remains the dominant system in use today. In this system, single-story-length studs are used to frame walls that support floor framing. Then, single-story studs are used to frame walls for the next floor that sit directly on top of the framing platform. Exterior sheathing in this period still relied on wood planks.

Materials. Plaster continued to be the primary finish for interior walls. Wallboard was invented in 1916 by the United States Gypsum Company. It was basically gypsum squeezed between two layers of paper. It was used extensively at the Chicago World's Fair in 1934 and had obvious installation advantages because it could be quickly nailed onto a frame with the seams plastered for a smooth wall. However, it failed to catch on during this time period.

Insulation. Insulation was not used in new homes during this time. Homes were smaller and advances in space heating technology had improved comfort, albeit with brute force rather than efficient thermal enclosures.

Windows. Most commonly, windows in new homes were single pane with wood frames.

Space Conditioning. Early coal-fired boilers and natural convection furnaces dominated home central heating until 1935 when the first forced air furnace was introduced. It used coal as a heat source and the power of an electric fan to distribute the heated air through ductwork. The electric fan continued to be the primary cooling technology available for American households. However, technology for modern air conditioning systems was beginning to be put in place. In 1902 Willis Carrier built the first air conditioner, but it was developed to combat humidity problems inside a printing company. In 1917 the first documented public building to use air conditioning made its debut at the New Empire Theatre in Montgomery, Alabama. Between 1928 and 1930, the chambers of the House of Representatives and the Senate, the White House, the Executive Office Building, and the Department of Commerce were air conditioned in Washington, D.C. By 1942, the nation's first "summer peaking" gas-fired power plant was built to accommodate the growing daytime electrical load from industrial and commercial air conditioning. However, residential air conditioning remained a luxury item for the wealthy until the post-World War II economic boom.

Plumbing. Plumbers used plain or tin-lined lead piping for cold water service, but they also had a choice of tin-lined, galvanized, enameled or rubber-coated wrought iron piping. Copper pipe was not added until after World War I. Storage water heaters continued to dominate the housing market primarily with electric and gas heaters. The main innovations with the storage water heater included better tank materials for improved durability, improved rust protection with glass lining and sacrificial anode rods used to protect the steel at imperfections in the lining, and added insulation around the tank to improve efficiency.

Electricity. Knob-and-tube wiring systems began to be phased out in the 1930s and replaced by nonmetallic and armored cable systems. Wiring continued to be made primarily of copper. Grounding requirements introduced by the National Electric Code in 1913 were typically achieved using the building's metal water piping as a grounding electrode. These requirements continued to be modified over this period. Plug fuses were used for overcurrent protection, but the clear window plug fuse replaced older opaque models, making it easy to see which fuse had opened. Circuit breakers resembling the modern systems used today were introduced in the 1930s.

U.S. Home Construction 1950–1985: Age of Cheap Energy

Production housing rapidly took off after World War II, with Levittown often recognized as the first prototypical suburban development. Of greatest significance is the emergence of insulated construction assemblies and air conditioning. The former was the precursor to building science in U.S. housing, and the latter was the technology that opened up southern states to massive development. Now that homes could be comfortably conditioned in high heat and humidity, the world would never be the same.

Foundations. Virtually all homes used poured concrete footings with either poured concrete or concrete block perimeter walls. Crawl space homes used wood piers on concrete footings for internal support.

Framing. Platform framing continued to dominate new construction, but the major innovation was the introduction of 4 ft by 8 ft sheet products for sheathing walls and roofs. The plywood production process was perfected in the mid-1930s, but World War II served as the proving ground for effective applications. By the late 1950s and early 1960s, plywood dominated all sheathing applications in the U.S. housing industry, and plank sheathing quickly disappeared. By the late 1970s and early 1980s, oriented strand board (OSB) was introduced as a lower cost alternative to plywood. However, it took about 20 years to get OSB right because the early product was excessively water sensitive (e.g., swelling edges). In addition, OSB is much more resistant to diffusion, which creates drying problems if wall assemblies get wet. This means you have to ensure rigorous attention to water managed walls and roofs.[12]

Materials. World War II proved to be the catalyst for mainstream use of gypsum wallboard. For the first time since the Civil War, there was a shortage of manpower. The old methods of plaster and lath were simply too cumbersome and manpower intensive. The U.S. government and various industries seemed to simultaneously discover the innovative sheet product that was given the name "drywall" because it went up dry rather than having to go on in wet plaster layers. The labor costs were indeed a fraction of lath and plaster. However, compared to lath and plaster, drywall made a hollow sound when knocked upon and was perceived as a low-quality finish. For this reason, many construction industry observers thought it was going to be a temporary replacement for the much more solid plaster wall system. But its low cost, square lines, and more uniform level surfaces proved too compelling, and the housing market quickly adopted drywall. Lath and plaster is now relegated mostly to historical renovation work.

Insulation. Insulation products were slowly introduced into housing after World War II, but they were not a high priority due to very low energy costs. Mineral wool and rock wool started to overtake asbestos in popularity in the 1940s and 1950s. Unfortunately, asbestos insulation products still made their way into hundreds of building products used in American homes between the 1950s and 1970s. This includes insulation for ventilation ducts,

[12] Joseph Lstiburek, "Building Science Insights, BSI-038: Mind the Gap, Eh!," February 22, 2010, http://www.buildingscience.com/documents/insights/bsi-o38-mind-the-gap-eh

plumbing, electrical systems, boilers, furnaces, and construction assemblies. The rediscovery in the mid-1970s of the harmful health effects signaled the end for asbestos materials in building construction. However, it is estimated that asbestos insulation is present in as many as 35 million U.S. homes today. Expensive safety regulations are now in place for its removal when it is identified in homes.

The desire for better-insulated homes changed dramatically on the heels of the oil embargos of 1974 and 1978, which led to a rapid escalation in energy costs. As a result, a large diversity of insulation products have been developed including batts (fiberglass or rock wool), blown-in products (cellulose and blown-in-blanket [BIB] fiberglass), spray-in products (open- and closed-cell polyurethane foams), board products (expanded polyurethane, extruded polyurethane, and polyisocyanurate), and advanced building systems integrated with insulation (structural insulated panels [SIPs] and insulated concrete forms [ICFs]).

Windows. In the 1950s, the technique of producing float glass was developed. Float glass is made with molten glass that "floats" over a tank of molten tin. It provides extremely flat surfaces, uniform thicknesses, and few if any visual distortions. Float glass is now used in virtually all residential windows. This proved to be a key breakthrough for developments to come such as low-e window coatings, which rely on the high-quality surface of float glass for the application of thin coatings. Until 1965, single-pane wood windows were most commonly used. The major innovation was combining single-glazed windows with attached storm windows and screens to reduce heat loss, especially for homes in colder parts of the country. About 1965, insulating glass was introduced (two panes of glass sealed together with an air space that functioned as an insulating layer) as an important step toward energy-efficient windows. Triple-pane windows were also introduced but only proved to be a niche product. Early product failures occurred with these multipane windows due to failed seals. By 1980, most of the seal problems were addressed and about 50 percent of the market had moved to double- or triple-glazed windows. As multipane windows evolved, the space between the layers was filled with argon or other inert gases to improve their insulating properties. Frames also continued to evolve. Manufacturers learned that aluminum frames were poor insulators, leading to significant heat loss. In the 1980s, manufacturers began to produce vinyl and wood-vinyl composite frames along with thermally broken aluminum frames for improved energy efficiency. At the same time, manufacturers began to replace the metal spacers typically used to hold multipane windows apart with foam or plastic spacers. Metal spacers were a source of heat loss that could often lead to condensation around the edges. In addition to improving energy efficiency, these nonmetallic spacers helped reduce window condensation.

Space Conditioning. Gas- and oil-fired versions of the old coal-fired forced air furnaces relieved homeowners from the chore of "stoking the coal fire" and rendered coal furnaces and cast iron radiators obsolete. But of even greater significance, air conditioning for the masses arrived in the early 1950s with the introduction of residential through-window and central air conditioning systems. An article on the history of space conditioning effectively expresses the enormous impact of this technology:

> *By 1953 room air conditioner sales would exceed one million units and by 1998 shipments of unitary air conditioners and heat pumps set a record of more than 6.2 million units. Unlike the impact of the relative convenience of central heating, air conditioning would have a profound influence on both building design and population migration and development. The air conditioner's widespread adoption would eliminate front porches, wide eaves and high ceilings from production housing and usher in the ranch house, "picture" windows, and sliding glass doors. Together, the inventions of central heating and air conditioning coupled with cheap and apparently abundant fossil fuels would free building designers from considering the external environment and allow them to use brute force heating and cooling solutions to overcome building designs totally inadequate for their local climates. Air conditioning alone would make possible the explosive post World War II growth of Sunbelt cities like Houston, Phoenix, Las Vegas, Atlanta, and Miami. . . . Air conditioning would also change our national patterns of living, turning us into 7/24 shoppers and gamblers trapped in giant malls and casinos without windows or any sense of time or place.*[13]

The following quotation from an exhibit on air conditioning at the National Building Museum in Washington, D.C., conveys the broader impact this technology has had on American life:

> *Before air conditioning, American life followed seasonal cycles determined by weather. Workers' productivity declined in direct proportion to the heat and humidity outside—on the hottest days, employees left work early and businesses shut their doors. Stores and theaters also closed down, unable to comfortably accommodate large groups of people in stifling interiors. Cities emptied in summers. . . . House and office buildings were designed to enhance natural cooling, and people spent days and evenings on porches and fire escapes. They cooled off by getting wet—opening up fire hydrants, going to the beach or diving into swimming holes.*

Plumbing. Copper with soldered fittings became popular around 1950 although it had been used as early as 1900. The solder used, 50 percent tin and 50 percent lead, is now understood to be a health hazard. This is due to significant leaching of the lead into potable water, particularly after long periods of low usage followed by peak demand periods.

Plastic supply piping has become increasingly common since the 1970s. However, some reports claim plastic pipes do not keep water as clean as copper and brass piping. This is because copper piping is bacteriostatic, meaning it doesn't support bacteria growth. In addition, durability issues developed with the first polybutylene plastic piping.

---

[13] "A Brief History of Heating and Cooling America's Homes," *Sustainable Dwelling,* October 26, 2007, http://www.sunhomedesign.wordpress.com/2007/10/26/a-brief-history-of-heating-and-cooling-americas-homes/

During this time period, the water heating business was faced with heavy competitive pressures that drove manufacturers to find ways to cut cost that compromised quality. This includes thinner steel, not offering double-glass lining, and replacing metal drains with plastic drains.[14] Solar water heating made a brief reappearance from the mid-1970s to the mid-1980s on the heels of significant tax credits. However with their high initial cost they could not survive the quick end to these credits.

Electricity. Homes built before 1965 were unlikely to have aluminum conductor branch circuit wiring. However, from the mid-1960s through the 1970s, aluminum wire became very popular because of the sudden escalating cost of copper. By the mid-1970s, significant cost reductions resulted in a quick market transition back to copper wiring. Aluminum is rarely used for residential branch circuit wiring today because of homeowner concerns surrounding earlier failures of aluminum wire. The National Electric Code introduced the first requirement for grounding type receptacles in 1947, but only for the laundry room. In 1956 the required use of grounding type receptacles was extended to basements. By 1962, the code was revised to require all branch circuits to include a grounding conductor or ground path for receptacles. That discontinued the use of nongrounding type receptacles in new construction. Accidental electrocutions in wet rooms were also a significant concern, with more than 1,100 deaths per year by 1970. The ground-fault circuit-interrupter (GFCI) was developed in the 1960s to address this problem. Starting in the early 1970s, GFCI receptacles were required by code in locations around the home where people would be standing on earth, cement ground, or near water. Lastly, the circuit breaker had almost completely replaced fuses by 1960 as the choice for overcurrent protection in new housing.

U.S. Housing Construction 1985–Present: Building Science Arrives

Buildings have become much tighter and better insulated since 1985. As a result, wall assemblies became colder and have less tolerance for drying (see Chapter 4) and the housing industry got a rude awakening to the risk of moisture problems and mold litigation. This has led to the emergence of building science as a critical new component of the housing industry essential to homes that work.

Foundations. Foundations predominantly rely on cast-in-place concrete although concrete block walls continue to be used for a much smaller percentage of homes. In addition, pre- and post-tensioned concrete was introduced in regions with expansive soils.

Framing. Due to depleting wood resources, the quality of milled dimensional wood framing is much reduced from years past. Wood framing has higher moisture content, more defects, and less dimensional stability. This reduction in quality has been visually obvious when

[14] Weingarten and Weingarten, "The History of Domestic Water Heating."

inspecting new home framing over the last 15 years. Engineered wood products have provided a significant solution for higher-quality framing. Options include finger-jointed studs, truss joists, laminated I-beams, and laminated veneer lumber (LVL) structural beams. Typically, engineered lumber is made from pieces of wood that are glued together for superior strength and dimensional integrity.

Materials. Housing materials in many markets went plastic: vinyl siding, vinyl trim, vinyl windows, plastic drain pipes, and plastic supply pipes. But many purists are worried about their "fake" appearance and poor "green" qualities since plastic is produced from petroleum by-products. In addition, composite products have made significant inroads, including cementitious siding, composite wood decking, and composite window frames.

Insulation. Diverse insulation products continue to be refined. High-density fiberglass provides greater R-value per inch. Spray foams have a new class of mid-density products (~.08 to 1.2 lbs). The SIP industry realized that panels need to be rigorously sealed at seams. ICFs have moved almost exclusively to plastic ties to minimize thermal bridging, and new larger ICF blocks were introduced to reduce installation costs.

Windows. Toward the end of the 1980s, low-emissivity (low-e) glass began to be incorporated in energy-efficient windows. The glass uses a thin layer of metal oxide to create a barrier to infrared radiation. Windows with this advanced technology coating work like magic, allowing reasonably high levels of visible light transmittance but blocking heat from entering in summer and escaping in winter. With double-pane glass, window mullions today are mostly decorative. Not only do they look fake, but they block daylight.

Space Conditioning. About 60 percent of U.S. homes are heated with gas-fired forced air furnaces, another 9 percent with oil-fired forced air furnaces, and about one-quarter of homes primarily in warmer climates use electric heat pumps for both heating and cooling. High-efficiency heating and cooling systems were introduced, including condensing furnaces and boilers, super efficient air conditioning and air-source heat pumps, and geothermal heat pumps.

Plumbing. By 1986, codes started to require lead-free solder for copper piping to address health concerns. Meanwhile, more durable CPVC and PEX plastic piping have become very popular with production builders and factory-built housing. Instantaneous water heaters and solar water heating are back, making significant inroads in the mainstream market. Instantaneous water heaters are being driven by significant improvements in the technology and strong market preference for their increased efficiency and "endless hot water" benefits. Solar water heating is being driven by a new round of tax incentives.

Electricity. Most developments revolve around continued refinements with electric wiring, outlets and switches, and circuit breakers. Structured wiring was introduced for information, telephone, and entertainment systems throughout the house, but wireless systems mitigate some of the need for hard wiring.

Why Quality Home Construction Is Broken

Key Lessons from Prominent Failures Have Not Been Learned

The corollary to stop protecting old technology is to not make critical mistakes adopting new technologies. Early failures can be devastating because it is a long road back once consumers and builders have a crisis of confidence. In addition, costs from mistakes with new technology can be enormous. Some of the most significant failures when adopting new technologies and practices over the past few decades are discussed in the following sections, along with these three key lessons learned:

- **Fully address durability issues**
- **Don't ignore the building science**
- **Check the ingredients**

Unfortunately, these lessons have not been integrated into the housing industry on a consistent basis.

Foundations

Cracked Foundations in Areas with Expansive Soils

Expansive soil is typically clay with "sponge-like" qualities that expands when exposed to moisture. In the worst locations, the soil can expand as much as 50 percent by volume, but 5 to 8 percent is most common. Even 5 percent expansion can exert 20,000 pounds per square foot of pressure on a home, causing floors and walls to crack, buckle, and heave. Expansive soil accounts for about 70 percent of foundation failures in the United States, with damage estimated at $4 billion nationwide. Locations particularly at risk include regions in Texas, California, Virginia, and Colorado. Colorado saw some particularly prominent failures because much of the Front Range sits atop expansive soils. Little was known about the problem until the 1960s because development was concentrated in central Denver with sandy soils that don't swell. Urban growth eventually pushed development to outlying areas with expansive soils, and major production builders were brought into class action suits throughout the Front Range by the 1990s.

Lesson Learned: Fully Address Durability Issues. Most builders could have avoided failures and lawsuits had they invested in soils testing. Soil maps showing locations of risk are widely available. Once an expansive soil condition is uncovered, it can be effectively addressed, albeit with a more expensive foundation. Typically slabs and homes are supported on foundation piers that extend to stable soil or bedrock. This is very cheap insurance for a builder.

Framing

Fire Retardant Plywood Roof Sheathing

Between 1985 and 1995, approximately 750,000 multifamily housing units experienced roofing problems due to a fire-retardant-treated (FRT) plywood sheathing failure. Although FRT was developed to increase resistance of plywood to flame spread during a fire, several brands experienced deterioration and severe strength loss within a few years after installation. It turned out that normally high attic temperatures would trigger their deterioration. These failures have led to nearly $2 billion in losses.

Lesson Learned: Fully Address Durability Issues. Because a builder is party to any product failure, new product innovations should not be employed until long-term performance is fully documented. When chemical compounds such as those developed for FRT plywood are introduced to the market, an investigation into product performance impacts should automatically be triggered.

Comfort and "Ghost" Mark Complaints with Steel Framing

Steel framing transfers heat and cold much more readily than wood, about 400 times faster. Thus steel studs typically lead to significant thermal bridges between indoors and outdoors. This has led to ghost marks on walls in winter where dust accumulates on colder drywall surfaces adjoining the studs. Comfort problems occurred due to colder surface temperatures in winter and warmer surface temperatures in summer. The visibly obvious defects often led to costly callback service and unsatisfied customers.

Lesson Learned: Don't Ignore Building Science. The fact that metals conduct heat incredibly faster than nonmetals is nothing new. Ignoring this basic science is a sure road to defects. There are lots of nice qualities to steel framing: it's renewable, low cost, easy to construct, much more dimensionally stable, moisture resistant, fire proof, and termite proof. Just make sure there is a complete rigid insulation thermal break wherever steel framing is used, including roof, wall, and floor framing exposed to unconditioned spaces. It's that simple. The most effective solution is to completely sheath all steel framing with at least 1 in. of rigid insulation. Some steel framing manufacturers are now making studs with integrated thermal breaks.

Materials (Exterior)

Hardboard Siding

Hardboard siding was developed in the 1920s as a way to use waste lumber from sawmills. Wood chips were heated in a sealed metal tube until they exploded, causing the wood to burst into fibers that were compressed with a steam press. The resulting material was more durable than plywood and serviceable. However, a cheaper, less dense process was adopted for hardboard manufacture in the 1980s. Homes with this product have experienced major dry rot failures with lawsuits starting in 1995. By the end of the decade, some major suppliers had abandoned the product. Today one manufacturer has redesigned the siding, correcting the original flaw in the production process.

Lesson Learned: Fully Address Durability Issues. When a product is new or formulations change, it's time to insist on long-term durability testing and empirical data. That said, even the old defective hardboard siding could have lasted much longer if the building science principle of bulk moisture management had been addressed. This would have entailed meticulous attention to detail, providing back and edge priming. Of course this is a challenging quality control issue, but it would have been much cheaper than the lawsuits and product replacement that ensued. And even with improved installation practices, owners would have had to be diligent sealing any breaks in the surface at nail heads, butt joints, and tears.

Manufactured Stone Veneer Moisture Failures

Over recent decades there has been a brewing epidemic of moisture failures with manufactured stone veneers used substantially in the western and south-central states. The author of one article reports investigating and repairing at least 100 recently built homes in his market.[15] The major problem is similar to that experienced with stucco walls. Both systems utilize a stucco basecoat that bonds to the building paper backing. As a result, water cannot drain, which can lead to moisture-damaged framing when excessive drying occurs to the inside. The problem with manufactured stone appears to be even worse than with stucco, including problems showing up sooner, progressing more quickly, and causing more severe damage inside the wall. This is attributed to a number of factors, including the greater thickness of manufactured stone that allows it to hold more moisture, ledge stone versions of the product that collect water and are often sloped to the wall, and the complication of practical flashing details due to the greater thickness.

Lesson Learned: Don't Ignore the Building Science. Water management is a critical principle of building science. At least two layers of backing are needed for stucco and manufactured stone, with the first acting as a bond break so water can drain at the second layer. In addition, all windows and doors require pan flashing details, and weep holes and flashing are critical at the bottom termination.

Materials (Interior)

FEMA Trailer Health Problems

Mobile home trailers are provided to disaster survivors as a helping hand, including to tens of thousands of Gulf Coast residents left homeless by Hurricane Katrina. Unfortunately, the inhabitants have been exposed to toxic formaldehyde gas emissions that can

---

[15] Dennis, McCoy, "Manufactured-Stone Nightmares," Impressions in Stone reprint of article in *Journal of Light Construction,* December 2004, http://www.impressionsinstone.biz/IIS-InstallationBestPractices.pdf

cause both immediate and long-term health risks. The formaldehyde gas emissions come from a wide variety of products used in these trailers, including composite wood and plywood panels.

Lesson Learned: Check the Ingredients. Some of the issues with urea formaldehyde foam insulation were unknown when used to retrofit homes with insulation in the 1970s. So using this product at that time was not unreasonable. By the time Katrina came along in 2005, however, the health issues associated with this dangerous compound were well known. Decision makers both at the manufacturing plants and FEMA should have known better and rejected products with formaldehyde. This would have been the much more profitable solution for everyone involved.

HUD-Code Manufactured Homes with Mold at Interior Walls

For a long time manufactured housing was the largest market-based segment of affordable housing. Large concentrations of these homes are found in the southeast and south-central states, which have long hot-humid seasons. Under these conditions, the prevailing driving force is from outside to inside: hot and humid air outdoors is driven to less hot and less humid conditions indoors. Yet the HUD code requires that exterior walls have a vapor barrier not greater than 1 perm (unit measure of vapor permeance) installed on the living space side of the wall. The HUD-code industry responded with vinyl covered wallboard as the typical interior finished surface.[16] This appears to have caused systemic failures in manufactured housing because wall assemblies could not dry to the inside. The vinyl finish functioned like a moisture dam. When moisture accumulations were large enough, mold could occur.

Lesson Learned: Don't Ignore the Building Science. The HUD-code industry was unfortunately given bad direction by the national code, but manufacturers should not have ignored the building science. Possible actions range from opposing the code with appropriate expert testimony to finding solutions that would have helped mitigate the risks involved (e.g., perforated vinyl finishes).

Chinese Drywall

During the housing boom after the devastating 2006 hurricane season, a severe shortage of drywall prompted many builders to seek out Chinese suppliers. As a result, more than 500 million pounds of possibly defective Chinese drywall was imported. According to the Associated Press, that was enough material to build around 100,000 homes. Chinese drywall was likely used throughout the country, but most of the drywall complaints have come from southern states where the warm, humid climate accelerates the emission of sulfur fumes. These sulfur fumes produce a "rotten eggs" odor and cause metals, such as air conditioning

[16] Neil Moyer, David Beal, David Chasar, Janet McIlvaine, Chuck Withers, and Subrato Chandra, "Moisture Problems in Manufactured Housing: Probable Cause and Cures," ASHRAE Conference Proceedings, Florida Solar Energy Center, 2001.

coils, to corrode. The fumes have also been associated with respiratory and sinus problems for some residents. Tests conducted by the state health department in Florida found that samples of Chinese drywall contained higher levels of sulfuric and organic compounds than an American-made sample. Lawsuits and settlements are still in progress but are sure to be a huge amount.

Lesson Learned: Check the Ingredients. When you choose a new technology, product, or source, ingredients should always be checked. This is especially important when the source is in a country known for lax regulations.

Insulation

Failed SIP Roofs in Juneau, Alaska

Structural insulating panel (SIP) construction is an integrated construction assembly that combines structure and insulation in one piece. This system provides thermal defect-free insulation, inherent air tightness, excellent structural performance, superior dimensional tolerances, and reduced construction time. Although this new technology offers these impressive benefits, it is not bullet-proof. SIPs were used for walls and roofs in a development in Juneau, Alaska, but poor quality control left an air gap at the ridge joint where roof panels meet. This ridge connection requires extra attention to detail because panels often have a much bigger gap than at vertical and horizontal edges, and the hygrothermal flow (moisture and air) is intensified there due to the stack effect and extremely cold weather conditions in Alaska. In this case, relatively humid indoor air found its way to that ridgeline gap and continually condensed on the cold outer OSB surface of the SIP roof panels over winter. Within six years the SIP roofs had rotted and needed to be replaced by the builder at great expense.

Lesson Learned: Don't Ignore Building Science. Here was an outstanding high-performance envelope system with one very costly building science mistake: the ridgeline was not fully sealed. It is critical to manage holes, particularly in cold climates. This is an especially silly mistake with SIP construction because the large panel sizes make it easy to visually inspect the substantially fewer cracks and holes than are found with conventional framing. Air will flow from more to less heat, humidity, and pressure and will take the path of least resistance through the largest available hole. The laws of physics are very predictable.

Urea Formaldehyde Insulation Health Problems

Urea formaldehyde foam insulation (UFFI) was manufactured in the 1950s, but it didn't become popular until the 1970s when energy prices began to rise. It was primarily used for retrofit applications because it could be more easily injected into existing wall assemblies with minimal holes compared to other insulation materials. Although an effective insulation, it poisoned homes by emitting formaldehyde during the curing process and for some time thereafter. The U.S. EPA recognized formaldehyde as a "probable" carcinogen, and major scientific reviews now link it to leukemia and have strengthened its ties to other forms of cancer.[17] Formaldehyde was banned for use in schools and residences by the United States Consumer Product Safety Commission in the early 1980s. However, much damage had been done. In 1977 the Congressional Research Service estimated there may be more than 200,000 installations that continue to require costly removal efforts across the country.

Lesson Learned: Check the Ingredients. Builders need to review the list of chemicals and materials used in every product specified. If they cannot do this themselves, they should hire an in-house building science expert or use a consultant on an as-needed basis. Today there are numerous formaldehyde-free insulation products to choose from. There is no need to take this kind of risk.

EIFS Homes with Dry Rot and Mold

The exterior insulation finish system (EIFS), sometimes called synthetic stucco, is an excellent technology for constructing a complete thermal break wall assembly with one big "if"—if you follow the manufacturer's specifications. Here is how the system works. A home is clad with rigid insulation and finished with lath and elastomeric stucco. Unlike traditional cement-based stucco, the EIFS elastomeric stucco is virtually impervious to moisture flow. However, hairline cracks between the finish and windows, doors, and trim eventually lead to moisture flow behind the stucco. In early applications of this technology, builders sought to save money and omitted the drainage layer specified in manufacturer installation instructions: building paper or house wrap under the insulation with pan flashing at windows and doors. The finish was impervious, so moisture flow through the cracks could not dry to the outside and was trapped inside the construction assembly. Dry rot became a problem when moisture accumulated at the wood assembly. In numerous reported cases, the damage was so severe 2 in. by 4 in. framing could be crumbled by hand. Leaving out the drainage plane was asking for trouble, and trouble obliged. At the peak of the crisis in 1995, an estimated 260 million sq ft of the product was applied with major problems reported both in high-humidity areas like the southeast and drier areas like Austin, Texas. This left many builders with costly class action lawsuits.

Lesson Learned: Don't Ignore Building Science. One basic rule of high-performance homes (see Chapter 4) is to effectively drain bulk moisture from roofs, walls, and foundations. It is extremely cheap insurance.

Windows

Lack of Pan Flashing at Windows and Doors

World-renowned building science expert Joe Lstiburek has many famous lines including, "there are two types of windows, those that leak, and those that will leak." His obvious point is that all windows leak. However, window and door pan flashing have been substantially

[17] Joaquin Sapien and ProPublica, "How Senator Vitter Battled the EPA over Formaldehyde's Link to Cancer," *Scientific American*, April 16, 2010, http://www.scientificamerican.com/article.cfm?id=vitter-formaldehyde-epa

Figure 5.2: Mold and Dry Rot Damage without Window Pan Flashing.

missing from home construction for decades, which means water is getting behind cladding and damaging homes all across the country. Figure 5.2 shows an example of the potential damage.

Lesson Learned: Don't Ignore the Building Science. Pan flashing at windows and doors is part of an obsessive focus on bulk moisture protection all builders should adopt. If not applied, it is a lost opportunity for the thirty-year or longer lifetime of windows. This is because it is cost prohibitive to go back and add flashing after construction is complete. This should be a no-brainer.

Space Conditioning

Early Heat Pump Comfort Failures

The electric power industry salivated over an opportunity to level the load factor on their power generation plants during winter months by providing heating with heat pump technology. Heat pumps were introduced in the early 1950s but suffered declines in the 1960s due to a record of poor reliability. However, rapid growth occurred after the 1970s when increasing electricity costs made electric furnaces less competitive. Utilities went on a full-court press to maximize the number of heat pump units installed, including in regions with colder climates. The major mistake was to ignore performance for the sake of number of units sold. According to a research report funded by EPA's ENERGY STAR program, more than 50 percent of all heat pumps have significant problems with low air flow, leaky ducts, and incorrect refrigerant charge.[18] These research results help explain the strong consumer preference for gas and oil space heating in moderate and cold climates.

Lesson Learned: Don't Ignore the Building Science. There's nothing wrong with electric heat pump technology. It just works differently and demands a strong companion emphasis on building science. Heat pumps do not have the brute force of conventional fossil fuel heating systems. For instance, conditioned air temperature for electric heat pumps is about 95° F compared to about 105° F for a gas furnace. Therefore, there is less tolerance for defects such as improper duct sizing, leaky ducts, and improper refrigerant charge. Efforts promoting this technology should have rigorously addressed complementary building science improvements such as properly installed insulation, comprehensive duct sealing, comprehensive air sealing, complete air barriers, high-performance windows, and quality HVAC installation. An effective thermal envelope and distribution system would have helped ensure customer satisfaction with heat pumps. By simply going for the numbers, the electric power industry invested billions of dollars convincing American homeowners that heat pumps are an inferior technology. Ignoring the building science substantially undermined their original goal to establish an important load-leveling technology.

Plumbing

Polybutylene Plastic Water Piping Failures

Polybutylene (PB) is a flexible, easy-to-cut gray plastic that is put together with simple crimp connectors. Introduced in the late 1970s, PB has been used in approximately 6 million homes in the United States. Unfortunately, there have been chronic leaks. Consumer complaints in Texas prompted the largest class action in U.S. history against the manufacturers of PB and resulted in a $750 million settlement. Lawsuits have been settled in other states as well. Manufacturers and other defenders of PB piping have insisted the product on the market today doesn't deserve its bad reputation, blaming the bulk of leaks and ruptures on improper installation, particularly at the joints. Manufacturers claim joint problems are addressed with an improved manifold design system. However, there are still questions about long-term durability, including whether signs of severe deterioration in plastic fittings or the pipe itself will show up in 10 to 15 years due to contact with oxidants normally found in public water supplies.[19]

Lesson Learned: Fully Address Durability Issues. PB piping was appealing because of its significant cost savings. New materials should be used only when there is absolute certainty of no increased risk. PB piping is no longer considered viable, but builders continue to use new CPVC and PEX plastic piping, which have proven to be much more reliable.

---

[18] U.S. Department of Energy, Energy Efficiency and Renewable Energy (EERE), "Energy Savers/Your Home," http://www.energysavers.gov/your_home/space_heating_cooling/index.cfm/mytopic=12620

[19] PropEx.com, "Classroom," www.propex.com/C_f_env_polybu.htm, 2002-2005

Electricity

Aluminum Wiring

Homes wired from the mid-1960s to the mid-1970s may have aluminum wiring. Since that time, aluminum wiring has been implicated in a number of house fires. These homes too often had defective wiring connections that needed to be retrofitted. Extensive wiring replacement can cost thousands of dollars, which has led some jurisdictions to no longer permit aluminum wiring in new installations. The housing industry has substantially returned to tried-and-true copper wiring.

Lesson Learned: Fully Address Durability Issues. Aluminum wiring when properly installed can be just as safe as copper wiring. However, when technologies are unforgiving of improper installation practices, builders should consider the risk too high if they are not willing to diligently employ necessary quality management solutions.

Technologies That Can Improve Quality Are Underutilized

At the peak of the American automobile industry crisis, Alex Molinaroli, president of Power Solutions at Johnson Controls, was aggressively pursuing new battery technologies that would enable a smooth transition to electric-powered vehicles. At that time he said, "The automotive companies' being in a crisis [means] now's the time for a disruptive technology. . . . The automotive industry has now moved beyond trying to protect old technology."[20] I hope Chapter 1 has made it clear that the housing industry has also arrived at a critical crisis, possibly one much more severe than that faced by the automobile industry. Certainly it is expected to be more protracted as the critical underpinnings (e.g., consumer confidence, accessible credit, and a healthy jobs market) continue to undermine a sustained recovery. Now is the time for the housing industry to move beyond trying to protect old technology. It currently takes too long for new technologies and practices to be adopted by the housing industry. This is particularly true for innovations that can solve real problems. This section looks at important new technologies and practices that should be considered but that remain off the radar screens of most mainstream builders.

Foundations. Apart from minor advances such as improved reinforcing and pre- and post-tensioning, traditional poured concrete foundation technology has experienced little change over the last century. Two important technology innovations are precast concrete and insulated concrete form (ICF) foundations. Both technologies have been on the market since the mid-1970s with very little traction. Why change foundation technology? The answer is a substantial improvement in quality and performance that should no longer be ignored. The bottom line is that basements utilizing either of these technologies cease to feel like cold, inferior living spaces, and instead are so comfortable, dry, and quiet they are fully comparable to above-grade construction. The value to the consumer should far offset the additional first cost.

---

[20] Fareed Zakaria, "To Pack A Real Punch, Everything Hangs on the Race to Build Tomorrow's Battery," *Newsweek,* February 21, 2009.

Figure 5.3: Precast Concrete Foundation System.

COURTESY OF SUPERIOR WALL CORPORATION

Precast concrete foundations are poured in a plant to precise dimensions and trucked to the site for installation (Figure 5.3). Impressive benefits accrue from engineered performance and precision work under controlled environment conditions:

- **The engineered 6 in. wall with a web configuration uses a third of the concrete of a typical 8 in. concrete wall while providing equal or better structural performance.**
- **Since 5,000 psi concrete is used rather than 2,500 psi for poured-in-place concrete, there is substantially greater resistance to water intrusion. Water control at panel connections is also enhanced with ship-lapped joints.**
- **Wood nailing strips are set at the edge of each web, providing a built-in nailing base for drywall with no further framing. By comparison, typical 8 in. concrete walls with 2 in. by 4 in. framed walls and a 1 in. gap consume many more square feet of available space. This can add up to about 70 sq ft in an average-size basement.**
- **The precast walls come with R-10 rigid insulation lining each web bay for a built-in code-compliant insulated wall assembly in most regions of the country. The insulation is totally bonded to the walls, so it effectively controls the surface temperature of the concrete year-round for dramatic improvements in comfort. Extra insulation can easily be installed in the cavities formed by the webs.**
- **The webs also include precast holes for running wiring with no extra work cutting holes.**

- **No footings are needed because the monolithic walls sit on a gravel base just like railroad tracks. This eliminates the cost for footings and enhances drainage from under the slab across to the perimeter drain tile.**
- **The walls are dead level, reducing rework on finishes and ensuring the floor above is also level.**
- **The whole integrated assembly goes up in a fraction of the time required for a custom formed and poured concrete foundation. Generally, precast panels are lifted into place with a crane in one day.**

The very low market penetration for this technology suggests the housing industry is slow to pick up on this impressive array of benefits.

ICF basements also offer impressive improvements in foundation technology. This technology uses blocks made of extruded polyurethane stacked like traditional concrete block. However, the configuration is reversed, with concrete poured into the voids after reinforcing is set. This results in profound performance advantages over traditional concrete walls by eliminating all thermal bridging with contiguous insulation on both the interior and exterior. In addition, the approximate R-18 insulation in this assembly is much greater than a typical insulated foundation wall. With no hard surfaces exposed, the basements are extremely quiet and comfortable. They cost much more than a conventional concrete foundation, but it's like getting another level of above-grade performance.

Framing. Although some regions of the country use concrete block construction (e.g., southern Florida), the U.S. housing industry has been substantially dependent on wood framing since the birth of the nation. However, only small incremental improvements have occurred over two centuries. This lack of innovation is compounded by a decreasing quality of wood for dimensional lumber as large mature trees become increasingly scarce. As a result, wood framing today has much higher moisture content, greater defects, and less dimensional stability than in years past. Meanwhile, important innovations in framing remain on the sidelines with barely niche market acceptance. These innovations include value engineered framing, structural insulated panels (SIPs), insulated concrete forms (ICFs) for above-grade walls, and modular home construction.

Value engineered framing uses wood more efficiently where it is needed for structural support (see Chapter 4). This reduces the cost of framing materials and increases insulation in the construction assemblies for improved performance. It does take more front-end planning to lay out ahead of time and coordinate locations of window and door openings, but automated drafting systems can minimize this cost.

Vented crawl spaces should no longer be used in any locations with long periods of hot and humid weather. Basic building science dictates that more-to-less driving forces will yield significant heat and moisture flow upward through the floor insulation where it can condense on the colder subfloor surface above. The other reason for moving away from vented crawl spaces is the substantial challenge effectively insulating floors. Insulation in floors has a natural tendency to fall away from the subfloor due to gravity (who knew?). If insulation is not aligned with the air barrier, its effectiveness is substantially undermined. The much preferred foundation system is an unvented crawl space where the dirt floors are fully covered with a capillary break (e.g., plastic liner taped at joints, piers, and walls) and insulation is installed on the vertical walls and band joists.[21] In other words, the crawl space becomes a mini-basement. This creates much more efficient space for running HVAC ducts, substantially improves year-round comfort in rooms above, and potentially provides extra storage space because it is dry and temperate.

Termite-resistant framing should be much more widely considered in high-risk areas. This would provide valuable peace of mind for homeowners who are forced to rely on costly pest services and potential exposure to dangerous chemicals. This is a virtual no-brainer for any area exposed to Formosan termites, now or in the reasonable future. This includes all Gulf states from Texas to South Carolina with at least one colony also found in California.

The Formosan termite is an invasive species that has been transported worldwide from its native range in southern China. A single colony may contain several million termites, compared to several hundred thousand for other subterranean species, and forage up to 300 ft. A single colony can consume as much as 13 ounces of wood a day and severely damage a structure in as little as three months. Once established, Formosan subterranean termites have never been eradicated from an area.[22] Are you scared yet? Homeowners sure are, even with conventional species. The basic technology options include framing that termites cannot eat, such as steel, concrete, or masonry. Preservative-treated lumber is also available, primarily with a borax-copper treatment often referred to as borates, which is considered environmentally benign.[23] One report indicated the treated lumber provided good protection from Formosan termites with Southern Pine, but was more variable with Douglas Fir. It concluded borax-copper treatments can provide adequate protection against Formosan termite damage in wood species or wood products that allow uniform distribution of the preservative.[24] In addition, certain wood species are naturally resistant to termite attack and offer an alternative to preservative-treated wood, but they are either protected species (e.g., Redwood) or not widely available for framing in the United States. Raised floor framing can provide protection by elevating the structure above the ground where it can be isolated from the termite moisture source and away from a colony's habitat. Yes these framing systems cost more, but think of the increased value of a new home substantially protected from termite risk compared to used homes in a high-risk area.

Structural insulated panels (SIPs) represent a completely different alternative to framing. Essentially they represent a technology shift from sticks to panels (see Chapter 4). The panels are engineered for outstanding structural and thermal performance that exceeds performance of

---

[21] Advanced Energy Corporation, "Buildings/Crawl Spaces," December 27, 2010, http://www.advancedenergy.org/buildings/knowledge_library/crawl_spaces/

[22] Wikipedia, "Formosan Subterranean Termite," September 6, 2010, http://en.wikipedia.org/wiki/Formosan_subterranean_termite

[23] Forintek Canada Corp., "Borate-Treated Wood for Construction—A Wood Protection Fact Sheet," 2002.

[24] Stan Lebow, Bessie Woodward, Douglas Crawford, and William Abbott, "Resistance of Borax-Copper Treated Wood in Aboveground Exposure to Attack by Formosan Subterranean Termites," United States Department of Agriculture, Research Note FPL-RN-0285, April 2005.

conventionally framed walls. They outperform framed walls thermally by eliminating all convective flow in the assembly with a solid core of polyurethane foam insulation that automatically includes complete air barriers fully aligned with the insulation. The walls are installed in a fraction of the time for conventional framing and are more dimensionally accurate for reduced rework on cracks and gaps. Also, fewer tools are needed on-site and there is substantially less waste. So many impressive advantages, but this technology can't seem to achieve meaningful acceptance. If the housing industry is not ready to build whole homes out of SIPs, they should immediately take advantage of "killer applications" where SIPs can solve real problems. In particular, they are well suited for the following applications where the housing industry currently experiences a high incidence of thermal defects: floors over unconditioned space such as garages and cantilevers, band joists, attic knee walls, and attic access panels.

Insulated concrete forms deliver high-performance above-grade walls as well as basement walls. In addition to the quality advantages already mentioned, they provide superior protection from severe climate conditions (e.g., tornadoes, hurricanes, high winds) and fire.

Modular homes are constructed in plants under controlled environment conditions. This allows for better protection of the materials, optimized ergonomics for more consistent workmanship, and quality control systems that can minimize defects. Like other innovations, there are opportunities for the housing industry to use this technology in a hybrid approach if it is not ready to embrace a paradigm change. Specifically, I believe production builders can team up with local and regional modular plants to provide finished kitchens and bathrooms. This concept is substantially enhanced when homes are designed with the kitchen and bathrooms around a central core "wet wall" (see Figure 3.10). This is a killer application for modular construction because finished kitchens and bathrooms are the most complex, fragile, and time-consuming elements of new site-built homes. Modular plants and their controlled environment settings can much more easily stock all the required cabinets, hardware, finishes, fixtures, and lighting and perform the involved plumbing, electrical, and finish work more efficiently, with more consistent quality, and less risk of damage. Thus, as proposed here, a production builder would build the much more field-friendly dry spaces and drop in one or more kitchen and bathroom cores delivered from a modular plant. In a fully mature business model, one could imagine home buyers shopping for their finished kitchens and bathrooms in showrooms much like they shop for new cars. This could take tremendous frustration and risk out of constructing homes on-site while yielding kitchens and bathrooms with higher quality finishes and workmanship at a lower cost.

Materials. Exterior cladding systems need much greater attention to positive drainage of bulk moisture in high-performance homes that have less tolerance for drying. Builders would realize substantial risk reduction by installing siding over furring strips. They provide an air space that ensures full drainage of any moisture that gets behind the cladding. Joe Lstiburek, a prominent building science expert, recommends at least 3/8 in. furring strips.[25]

[25] Lstiburek, "Building Science Insights."

In very cold climates, builders committed to an inside vapor barrier should not use standard plastic lining under drywall. Cold climates in the United States can still have hot and often humid summers when driving forces will result in air flow from outside to inside. Thus, for a few months during the year, any plastic lining will function as a dam that stops construction assemblies from drying to the inside. Smart membranes are available that look like plastic but change permeability and become more porous as humidity rises. This provides a perfect solution because interior moisture is blocked during dry winter conditions, but exterior moisture is allowed to dry to the inside during hot, humid summer conditions. Eventually builders and code officials will figure out that plastic vapor barriers are not needed even in cold climates because air-tight drywall construction with a semivapor impermeable paint will provide more than adequate control of moisture diffusion.

Insulation. Spray foam is much more expensive insulation, but it solves real problems that cause common defects in new construction. In particular, preferred low-cost insulations used today are difficult to install in floors over unconditioned spaces (e.g., bedrooms above garages and cantilevers), band joists, and for sloped roofs in unvented attics. Conventional batt insulation products commonly result in installations with gaps, voids, compression, and misalignment and missing air barriers that undermine their effectiveness (see Chapter 4). In contrast, spray foam delivers three for one: insulation that inherently goes in without gaps, voids, compression, or misalignment, air sealing of any remaining cracks and holes, and an air barrier because the insulation itself resists air flow. Note that spray foam can still experience defects if it is not applied by a properly trained applicator under specified conditions (e.g., surfaces should not be wet).

Rigid insulated sheathing creates a completely thermally buffered wall while shifting the dew point outside the construction assembly (see Chapter 4). Where structural sheathing is needed, structural insulated sheathing (SIS) is a new product that combines rigid insulation with thin structural sheathing that enables walls to withstand 100 mph winds.

Advanced wall systems such as SIPs and ICFs discussed previously also provide superior insulation technology options for new homes.

Windows. Some of the best windows on the market today deliver solar heat gain coefficient and U-values at and below .30. In addition, super high-performance dual-glazed windows are available with a solar heat gain coefficient at and below .25 (blocking at least 75 percent of incident solar heat gain) and U-values at and below .15 (R-value of 7 or greater). New triple-glazed low-e superwindows are being developed that promise to deliver a U-value as low as .1. At this level of performance, some research analysis indicates windows can outperform the insulated wall.[26] In addition to improved glazing, some windows feature advanced materials including composite frames much better matched to the coefficient of expansion of wood framed walls than plastic or metal frames. This will minimize problems where different expansion coefficients lead to gaps between windows and walls.

---

[26] Darius Arasteh, Stephen Selkowitz, and John Harmann, "Detailed Thermal Performance Data on Conventional and Highly Insulating Window Systems," ASHRAE/DOE/BTECC Conference, December 2–5, 1985.

Figure 5.4: Operational Shutters in Italy.

While it's exciting to see windows get this good, a trip to Italy a few years ago reminded me of an old technology that performs even better blocking the sun in hot climates. Virtually every home had operational shutters that effectively provide a solar heat gain coefficient of zero (Figure 5.4). In other words, shutters in Italy are not just for decoration. What a concept. With higher costs of energy and possibly a greater social consciousness about wasting energy, Italians are highly disciplined about opening and closing their shutters each day. If this is perceived as too difficult for American households, the housing industry should be considering automated window shades for hot climates (Figure 5.5).

Space Conditioning. The most immediate need for space conditioning in the housing industry is to stop the madness of locating heating and cooling ducts in unconditioned attics and crawl spaces. Air heated to 105° F for space heating should not be distributed in attics that are below freezing. Air cooled to 55° F for space cooling should not be distributed in attics that can reach 140° F and hotter. As mentioned earlier, conditioned air experiences 10 times greater heat loss when distributed through unconditioned rather than conditioned spaces. The easy fix is to run ducts through floor truss joists selected with adequate space or to use dropped ceilings in hallways as a duct plenum. Alternatively, attics can be insulated at the sloped rafters rather than the ceiling joists, which creates a conditioned attic. Not only is this much more efficient for HVAC distribution, the tempered space is now suitable for additional storage needs. Note that high-performance homes should also use much smaller

Figure 5.5: Automated Window Shades.

compact duct systems because advanced wall assemblies and windows no longer need supply air distributed to the extremities. This was necessary in older homes that were leaky and poorly insulated to control surface temperatures and minimize window condensation.

The next technology step for space conditioning is to use more efficient HVAC systems with modulating air handler fans and air-tight air handler cabinets. Modulating air handler units are equipped with variable speed DC motor fans that are much more efficient than single-speed AC fans. This also allows HVAC systems to do double duty as distribution systems for fresh air ventilation without a significant energy penalty. An air-tight air handler cabinet is an important efficiency enhancement because as much as half of the leakage in a typical duct system occurs in the cabinet. A number of manufacturers are now providing air-tight cabinets.

Another innovation currently being developed by the HVAC industry is integration of whole-house dehumidification with air conditioner and heat pump systems. This is an important technology improvement for hot and humid climates where large latent loads persist during swing seasons when there are no or minimal sensible cooling loads. This technology will have impressive benefits both for energy savings and indoor air quality through better moisture control.

Geothermal heat pumps are a highly efficient option for all-electric space conditioning. Rather than using outdoor air as the source of heat exchange, they use the ground. Their efficiency advantage comes from more moderate year-round ground temperatures below the frost line at approximately 55° F. Compared to air temperatures, this is much cooler in summer and warmer in winter. The rub is that the vertical or horizontal ground loops (e.g., heat exchangers) are much more expensive than an outdoor heat exchanger used in air-source heat pumps. Geothermal heat pumps are well matched to high-performance homes that substantially minimize space conditioning loads and in turn minimize the size of the required ground loop. Further cost reductions for this technology will be possible with economies of scale and advances in ground loops. But why should builders be interested in this new technology? First, it eliminates the need for an outdoor compressor. This makes homes and neighborhoods much quieter, which is often one of the most desired benefits. In fact, dense developments in hot climates often resemble "compressor farms" with the ugly and noisy boxes outside every home. With only one internal unit, there is no need for a field-installed refrigerant line from the outdoors to the internal unit. Most geothermal heat pumps are hermetically sealed with refrigerant in the factory for improved quality installations. In addition, the lifetime of geothermal units without an outside compressor is much greater than air-source heat pumps exposed to harsh weather and sometimes corrosive conditions (e.g., salt air in coastal regions). Finally, properly installed geothermal heat pumps can provide significant energy savings.

Plumbing. Water heating has long been the unassailable load without many options for improved performance. However, instantaneous water heaters and solar water heating are now making significant inroads, and both contribute significant energy savings. Instantaneous water heaters generally can improve the energy factor efficiency from .6 for a conventional high-efficiency storage tank water heater to above .85. An important secondary benefit for instantaneous water heaters is the endless supply of hot water without a tank. Solar water heaters typically entail a solar collector that heats water either stored directly behind the glass or circulating in pipes through the collector. Various technology options use both active and passive water circulation and offer different levels of freeze protection. These systems generally can provide solar fractions ranging from .5 to .9, which means 50 to 90 percent of all the water heating energy comes from the sun. Solar water heating should no longer be ignored in hot and moderate climates with high levels of solar insolation.

Once hot water is produced efficiently or renewably, efficient hot water distribution is the next important water heating technology innovation ready for prime time. The first option is the design solution discussed earlier called core plumbing. It relies on the house designer to develop a floor plan with all wet rooms off a central wet wall (see Figure 3.11).

The other option is structured pumping (Figure 5.6), which does not require any design accommodations. Additional piping is provided as required to form a complete cold water piping loop from the most remote fixture back to the hot water heater. A pump is located in the loop, typically next to the heater. Switches or occupancy sensors at each fixture are used to activate the pump, which distributes hot water until a sensor detects hot water has been fully distributed and sends a signal to shut off the pump. Structured pumping systems provide a number of important benefits. Nearly instant hot water is provided to faucets, which is a performance advantage highly valued by homeowners. Energy can be saved, particularly during colder months, because water in the pipes returned to the tank is generally at room temperature,

Figure 5.6: Structured Plumbing Diagram.

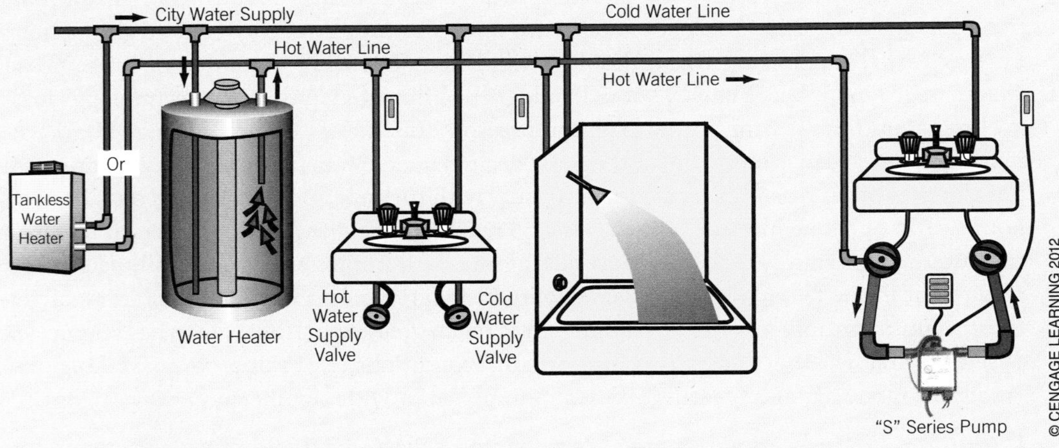

Source: Courtesy ACT Metlund® D'MAND® Systems.

which is much warmer than water from the street main. And since it is no longer necessary to run water down the drain waiting for hot water, significant amounts of water are conserved. These savings can amount to as much as 5,000 to 7,000 gallons a year for a typical household.[27] An additional benefit is longer life for storage water heaters because circulating water helps to keep the water mixed so less sediment accumulates at the bottom that can decay the tank.

Another important best practice substantially ignored by the housing industry is to carefully plan plumbing layouts. This entails basic concepts for efficient plumbing distribution and proper pipe sizing that can optimize the delivery of water, reduce water consumption, and save material.[28]

Electricity. The second wave of new home buyers almost as large as the 76 million baby boomers are the 72 million Gen Yers currently between the ages of 17 and 34. By 2014, they will represent the largest population of new home buyers. This generation loves technology and connectivity. The connected home is happening one network at a time for security, data, and entertainment. Home area networks (HANs) are already making their way into homes for managing all computer, printing, and phone operations. Residential energy management (REM) systems are increasingly available that enable automation and control of space conditioning, entertainment, security, broadband access, entertainment, and lighting. Often this control is centralized with home computers and accessible with cell phones. Currently,

[27] Larry Aker, owner of ACT Metlund D'Mand Pumping Systems, personal communication, August 16, 2010.

[28] Gary Klein, Managing Director of Affiliated International Management, LLC, personal communication, March 24, 2010.

REMs are associated with utility "smart grid" deployments, but opportunities will not always be utility driven. When connected to smart grid applications, these systems will also enable load control modules for demand response savings, micro-generation management for on-site power systems such as solar electric and micro-cogeneration, and energy storage and retrieval for forthcoming electric vehicles. In addition, these systems will provide a "dashboard" interface much like in new cars that will allow owners to better track and manage their energy consumption. Old timers will be overwhelmed; the new generation will love it. Be ready.

Another major technology innovation about to emerge is solid state lighting, also known as LED (low emitting diode) technology from computer chips. Not only does this technology provide the same or better energy savings opportunities as current compact fluorescent lighting, it is inherently adaptable to full dimming, lasts much longer, and doesn't utilize mercury, eliminating the associated hazard exposure risk and disposal burden. Look for all new homes to be substantially lit with this advanced new lighting technology within the next 5 to 10 years.

The housing industry also needs to address the increased interest in net zero homes. These homes offset on-site electricity consumption with on-site electricity generation through solar electric systems. The interest in this technology has been substantially driven by large tax incentives and utility rebates, but increasing energy costs and future cost reductions are likely to sustain growth. Builders not ready to dive in now with solar electric systems should offer solar-ready homes. Typically these homes include enough south-facing roof area (+/– 45 degrees off true south), additional structural capabilities needed for solar panels, conduit to run electric wiring from the roof to a power inverter, a location set aside for the inverter, and modifications to the electric circuit breaker and meter. Marketing a solar-ready home can be a significant advantage for a home builder.

A similar concept is adding features to make a home electric-vehicle ready. Many automobile industry experts are predicting a large penetration of electric powered vehicles as manufacturers begin introducing plug-in hybrid and all electric cars. This entails setting up a 220-volt charging outlet in the garage, much like the one for a clothes dryer or central air conditioner. Normal household power will charge some plug-in hybrids (e.g., Chevrolet's Volt), but the 220-volt outlet requires only half the charging time. In addition, a so-called smart outlet would charge at night when power usage and prices are lowest with time-of-use electric rates. KB Homes has already jumped on this innovation and has received a lot of press attention for offering this simple modification to their homes.

Quality Assurance Processes Are Not Applied

Until World War II, U.S. housing was constructed by craftsmen builders who did most of their own work assembling structures and finishes custom fit for each job. The origin of modern manufacturing processes is often attributed to Henry Ford for his system of interchangeable parts and mass production used to manufacture the Model T in 1913. It took the advent of the suburban development following World War II for some components of mass production to make their way into mainstream housing. However, critical components of quality systems are absent. John Tooley, a renowned building science expert, speaks across the country on the quality systems in home construction citing that "process fails more than workers!"[29] Tooley states "it is better to have common people managing a superior process than to have superior people managing a bad process." Quality assurance processes are definitely broken in the housing industry based on the following observations:

Selection of materials and practices. The housing industry is protecting old technology and missing the opportunity to employ many profoundly better materials and practices. Moreover, the specification of new technologies in the past has not consistently applied building science, ensured durability, and carefully checked the ingredients.

Documentation. Contract documents are often prepared only to meet minimum requirements for securing building department approval. As a result, many quality construction and building science details are routinely absent.

Checklists. Detailed checklists, often under the name "QC traveler" (a.k.a. quality control checklist), are used at every step of the assembly process in production plants for HUD-code and modular homes. However, these factory-made homes represent a very small percentage of new homes constructed each year. Checklists, a basic quality assurance tool, are conspicuously missing in field production for site-built homes. Yet virtually every other industry employs detailed checklists and diagnostic procedures. These are missing in site-built housing leading to an industry with excessive callback problems and service center costs.

Training. Training has been a low priority in the housing industry when it comes to preparing trades for new innovations in building science and best installation practices.

Testing and inspections. A classic line about quality assurance is that "you can't manage what you don't measure." Before many of the recent building science programs, there were no routine measurements of air leakage, duct leakage, quality of insulation installation, completeness of air barriers, proper installation factors associated with heating and cooling systems, and effectiveness of the water management system. Every builder should routinely employ these and other testing protocols.

Root cause analysis. Total quality management (TQM) and its more recent manifestation as Six Sigma both emphasize minimum defects through optimizing processes and resolving problems at their root cause. One of the techniques to come out of these management systems is Kaizen. It utilizes single events to solve critical problems and can be applied as an ongoing tool for continual improvement. Multidisciplinary teams are assembled sometimes with outside stakeholders to brainstorm solutions and innovations. The housing industry needs a stronger commitment to this type of root cause problem solving. Simple cosmetic changes will no longer be enough to support a successful home building business.

---

[29] John Tooley, "Mixing the Quality Serum," presentation at RESNET Annual Conference 2011, Orlando, Florida, March 2, 2011.

Lean Production Processes Are Not Consistently Utilized

My personal observations of the housing industry are that lean production systems are far from fully integrated. Here is a short list of the worst offenders:

> ***Overproduction.*** Houses are far larger than needed to meet consumer special needs and often have lots of wasted space because furniture layouts have not been carefully planned and built-in furniture is underutilized.
>
> ***Overprocessing.*** Homes today still have too much framing waste, oversized HVAC systems, excessive duct layouts when compact layouts work just as well, inefficient plumbing layouts, and lack of overall system integration that could minimize waste.
>
> ***Correction.*** Quality control systems are underutilized, resulting in excessive customer service costs for too many builders. Full integration of all systems is poorly addressed leading to costly and suboptimal solutions in the field by different trades.

How to Fix Quality Home Construction

Competing on price is a weak hand for builders in a market overwhelmed with "fire-sale" prices driven by foreclosures, short sales, and crisis sales. The following retooling recommendations require investments in quality construction that will take housing to an entirely new level of excellence. This is in fact an opportunity to make a compelling case for purchasing new rather than used homes. The home retrofit industry will realize commensurate benefits as superior performance in new homes affects aspirations and demand for home improvement projects. A short-sighted first-cost business model has kept the housing industry down for too long. It's time to make the following changes.

Invest in Comprehensive Construction Documents

Another term for construction documents is "contract" documents because they form a legal framework for the materials and services provided by all contractors working on a project. Construction documents prepared with the minimum effort needed to secure building department approval will not get the job done. Details and specifications should clearly define workmanship and materials that will hold all subcontractors legally responsible to quality standards. At a minimum, construction documents should address the following criteria:

- **Comprehensive building science measures**
- **Specifications for materials whose ingredients and durability have been verified to minimize risk**
- **New technologies and practices that can ensure optimum performance**

Do these criteria sound familiar? They are the recurring reasons for failures experienced by the housing industry that were highlighted earlier. Getting the details right on the contract documents is the critical first step. At a minimum, construction documents should include the following information:

- **Detailed structural framing layouts that specify only wood needed for structural purposes to maximize insulation and reduce cost**
- **Detailed HVAC system layouts that include compact duct layouts inside the conditioned space with necessary details to ensure adequate space and clearances**
- **Details and specifications for air-tight duct assemblies and air handler cabinets**
- **Complete specifications for quality installation of HVAC equipment that include sizing of equipment, ducts, and terminals; air distribution; matching components; and refrigerant charge**
- **Details for HVAC filtration that effectively removes particulates to 3 microns or less and is properly air sealed inside the duct assembly**
- **Thermal enclosure system details that include comprehensive air sealing measures, air barrier assemblies, proper installation of insulation, type of insulation optimized for each application, minimized thermal bridging measures, and advanced window specifications**
- **Detailed wall and whole-house sections that clearly delineate boundaries between conditioned and unconditioned spaces**
- **Complete roof, wall, and foundation water management details, including heavy membrane specifications at valleys and eaves; complete weather-resistant barrier at all walls; flashing details at wall–roof junctions, exposed roof underlayment, siding termination, windows, and doors; drain tile with fabric filter and capillary breaks at slab and footings**
- **Material selection and specifications that ensure protection of stored materials during construction and appropriate materials for wet areas**
- **Complete lighting design, layouts, and fixture specifications**
- **Complete plumbing layouts that minimize piping and maximize efficient distribution of hot water**

Invest in a Building Science Manager

From personal experience providing training to thousands of builders, it is disturbing how few employ an in-house building science expert. Claiming ignorance just doesn't work. The housing industry has made too many costly mistakes by failing to address building science.

With the advent of building science programs, many builders are using Home Energy Rating System (HERS) raters to verify that their homes meet guidelines. However, building science is far too important to relegate to end-of-construction inspections. Building science principles explain why homes work and fail, so it stands to reason that *every* builder should have a building science manager in-house. This can be the owner, an officer of the company, or an in-house expert assigned to that position. A weak second option is to have a building science expert on retainer for continuous consultation.

The primary responsibilities of the building science manager include ensuring all design decisions, products specified, and scopes of work address comprehensive building science, pollutant control, and fortified construction; hiring trades qualified to consistently achieve building science objectives; and performing spot field inspections and diagnostics needed to hold subcontractors accountable. This manager must be empowered to impose product and process changes needed to enforce building science principles. Large national builders should also have a national building science manager to coordinate proven solutions across the entire product line.

This retooling recommendation needs to extend beyond home builders; it is time for the architectural profession to also embrace building science as a critical core competency. This begins by setting curriculum requirements for all accredited university programs and even vocational schools to incorporate extensive building science courses and require full application of building science in design classes. It's amazing what I did not know about why buildings worked and failed after graduating from architectural school. In talking to recent graduates some 35 years later, my sense is that not much has changed at far too many schools.

Invest in New Technologies and Practices

Risk adversity to new technologies and practices is understandable when viewed against the prominent and costly failures discussed earlier. However, builders must build past this reticence and actively consider a long list of impressive new technologies and practices that solve real problems or provide opportunities to improve home performance. Some of these have been discussed earlier in this chapter. Builders cannot afford to sit on the sidelines any longer, but the following selection criteria should be rigorously applied:

- ***Don't ignore the building science.*** Many of the housing industry failures discussed in this chapter could have been avoided if business decisions applied the principles of air, thermal, and moisture flow. This is one reason it is critical to have an in-house building science manager or consultant.
- ***Read the ingredients.*** Every material and chemical used in a new product or in an existing product from a new source should be analyzed for health and performance issues. As a builder, you are fully liable. If there are any uncertainties, experts should be consulted. A product should never be used to reduce cost that will increase the risk of occupant health impacts or product failure.

- ***Verify long-term durability.*** New products or practices should not be employed without a full durability assessment. At a minimum, this assessment should include documentation of long-term performance or expert analysis ensuring materials and assemblies entail no, or minimal, risk.

Invest in Quality Assurance Processes

Prior investments in construction documents and a building science manager or consultant set the stage for implementing the following comprehensive quality control processes:

- ***Secure top down commitment.*** Unless the head of the organization is totally dedicated to quality processes, operations will revert to form under pressure. And there is plenty of pressure in the home building business.
- ***Develop detailed scopes of work.*** Comprehensive construction documents will provide actual details and specifications. However, they must be a continually moving target as innovations offer opportunities for less waste and better performance.
- ***Train to detailed scopes of work.*** Training needs to be an ongoing activity to ensure all trades and in-house staff are prepared to successfully conform to the scopes of work. It is critical that all trades be engaged as a team to coordinate and collaborate on optimum construction processes and efficient use of materials.
- ***Measure performance.*** Detailed checklists, critical diagnostic tests, and other visual inspections should be formally reported to all levels of the company. Opportunities for improvement and innovation will be revealed by these measurements.
- ***Address root cause solutions for nonconformity.*** Where there is nonconforming work, engage in root cause analysis rather than just addressing symptoms. Often this will identify processes that need to be fixed, materials that need to be changed, or workers and subcontractors that need to be held accountable.
- ***Continually improve.*** There are lots of systems for continual improvement. Builders should select and implement a proactive process to innovate their product to minimize defects, maximize performance, and optimize aesthetic appeal. Moreover, this process should engage all internal and external stakeholders vested in the final product.

Invest in Lean Construction Practices

I have attended numerous presentations by experts in lean production who provide extensive examples of impressive reductions in cost and risk when comprehensive lean production techniques are applied. Builders need to make significant investments in experts and in-house systems needed to employ lean construction on a broad scale. This will typically entail coordination with designers, construction management staff, all trades, and suppliers because home production is a very complex process.

Chapter 5 Review

So What's the Story?

Housing construction has experienced little change and lots of change over time. Homes are still primarily built with sticks, but today we live better than royalty in past centuries with running hot water on demand and year-round comfort control. The quality home construction story can be summarized as follows:

- **What It Is.** Application of the right technologies and practices with minimum defects and waste.
- **How It Got Here.** Centuries of development have incrementally advanced all systems in home construction.
- **Why It's Broken.** The housing industry has not learned critical lessons from prominent failures and remains reluctant to use the best available new technologies and practices.
- **How To Fix It.** Top-down commitment is needed to invest in all facets of quality home construction, including adopting new technologies and practices, instituting comprehensive quality assurance processes, and employing lean construction.

The details of this story are included in Table 5.3.

Table 5.3: Quality Home Construction Summary

What It Is	How It Got Here						Why It's Broken	How To Fix It
	Ancient Origins	1750–1850	1850–1920	1920–1950	1950–1985	1985–Present		
Right technologies and practices							Durability not addressed Building science principles ignored Ingredients not tested	Use lean construction methods Provide detailed contract documents Adopt new technologies and practices
Foundation	Egyptian pyramids Romans 100 BC	Stone or brick walls	Stone or brick walls	Poured concrete Concrete block	Poured concrete or concrete block	Post-tensioned concrete for bad soil	Cracked foundations with expansive soils	Precast concrete ICFs
Framing	Chinese timber frames 770–446 BC	Timber framing	Framed cavity with balloon framing Plank sheathing	Platform framing replaces balloon framing Plank sheathing	Platform framing 4 ft x 8 ft sheet plywood OSB replace planks	Engineered lumber Advanced walls SIPs ICFs	Fire retardant plywood sheathing Steel framing ghosting	Value engineering Framing SIPs ICFs Termite resistance

Materials	Masonry and stucco used by ancient Greeks and Romans	Wood and masonry siding Exposed framing	Wood and masonry siding Plaster	Wood and masonry siding Plaster	Wood, masonry, stucco, and aluminum siding Drywall	Wood, masonry, stucco, and vinyl siding Drywall	Hardboard siding Manufactured stone FEMA trailer formaldehyde HUD-code vinyl walls	Air gap between siding and weather resistant barrier
Insulation	Asbestos with ancient Greeks Tapestries used in Middle Ages	None	Open framed walls Asbestos around steam pipes	Open framed walls	Insulation becomes common in 1970s Asbestos still used in many products	Diversified products: Batt Blown-in Spray foam Rigid board SIPs ICFs	SIPs in Juneau, Alaska Urea formaldehyde insulation EIFS moisture problems	Spray foam Rigid insulating board sheathing SIPs ICFs
Windows	Smoke hole First transparent window in ancient Rome	Cast plate glass only in small sizes	Cast plate glass available in larger sizes	Single-pane windows in wood frames	Float glass developed Multipane windows	Low-e windows Advanced frames and spacers	Lack of pan flashing	Exterior shading devices Operable shutters Solar orientation

(*Continued*)

Table 5.3: Quality Home Construction Summary (Concluded)

What It Is	How It Got Here						Why It's Broken	How To Fix It
	Ancient Origins	1750–1850	1850–1920	1920–1950	1950–1985	1985–Present		
Space conditioning	Indoor fires Natural comfort with south orientation Thermal mass Cross ventilation	Burning wood in fireplaces and cast iron stoves Natural comfort	Coal supplants wood-fired heating in 1885 (boilers and furnaces) with natural convection Natural comfort	Forced air coal furnace AC introduced as luxury item	Gas- and oil-fired furnaces replace coal AC for the masses arrives	High-efficiency furnaces, boilers AC standard feature	Early heat pump comfort problems	Ducts in conditioned spaces Compact ducts Variable speed air handler AH Air-tight air handler AH Natural comfort
Plumbing	First flushing water closet appeared 2,800 years ago in Crete Aqueduct system in ancient Rome provides water for indoor plumbing and baths	Hand pump and pail Outhouses Some waterworks projects	Waste water, indoor plumbing and water closets make inroads Advanced water supply systems Water heating introduced	Mostly tin-lined lead or wrought iron pipe, but copper piping introduced after WWI Storage water heaters dominate market but improve	Copper piping with soldered fittings Plastic piping Solar water heating briefly available	Lead-free solder More durable plastic piping (CPVC and PEX)	PB plastic water pipe failures	Core plumbing design Structured plumbing Efficient plumbing layouts Solar water heating

Electricity	None	None	Knob-and-tube electricity introduced	Nonmetallic and armored cable copper wiring replace knob-and-tube, plug fuses	Aluminum wiring introduced briefly Grounding type receptacles, circuit breakers replace fuses	Wiring, outlets, switches, circuit breakers refined Structured wiring	Aluminum wiring fires	Control systems LED lighting Solar ready Electric-vehicle ready
Ensure work done right	Craftsman builders	Craftsman builders	Craftsman builders	Craftsman builders	Production builders don't apply quality assurance processes	Expands into other U.S. industries	Outsourced contractors not effectively managed	Invest in quality assurance processes
Ensure minimal waste and continual improvement					Principles developed by Toyota Motor Co.	Introduced in 1992 in construction	Excessive waste in the production process	Invest in lean production

Retooling THE U.S. Housing Industry

6

Effective Home Sales: Better Is Not Good Enough

Effective Home Sales: Process, Goals, and How Goals Are Achieved

Process. *Purchasing a home can be an overwhelming experience, evoking soul-searching considerations from basic nesting instincts to heartfelt aspirations. It is a very complex transaction process and not for the faint of heart. Sales is the final, and perhaps the most important, component for the housing industry. If you cannot sell your product, there is no business. The sales process entails a wide range of marketing efforts to get potential home buyers interested in your product. Until recently, signage was the number one marketing tool for home builders. Today the Internet has emerged as a key "touch point" for American home buyers: more than 80 percent search the Internet before purchasing a home. However, traditional advertising in local newspaper real estate sections, real estate journals, and radio continue to be used by many builders. Once a home buyer contacts a builder, the formal sales process begins. A real estate agent, in-house sales agent, or builder typically works directly with the prospective buyer. When a purchase decision is made, contract documents are drawn up with financing and the real estate closing process concluding the transaction. After closing, the buyer is given title and keys. Thereafter, there is a warranty period during which the builder is obligated to address customer concerns and make necessary repairs.*

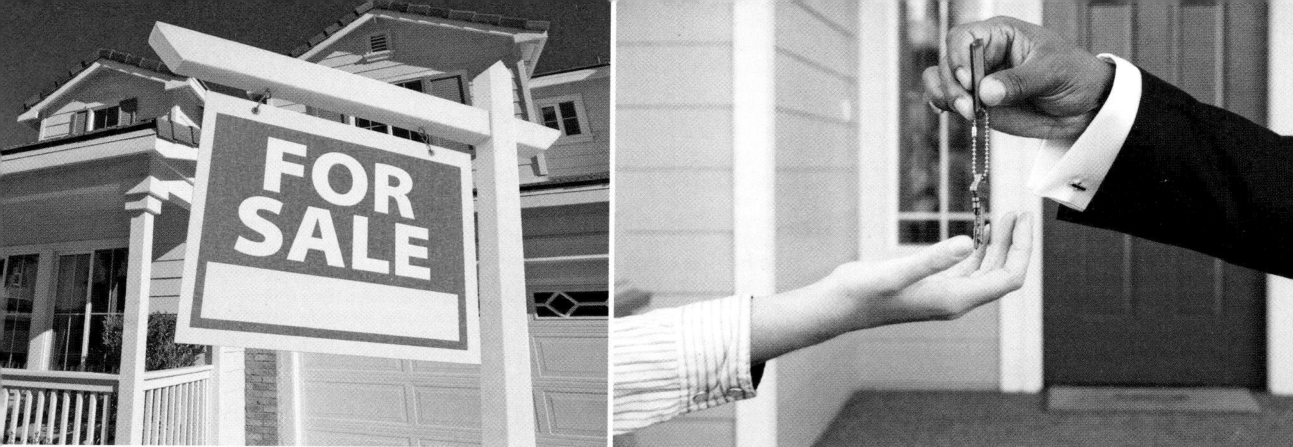

© iStockphoto.com/Feverpitched © iStockphoto.com/akurtz

Goals. *The first goal of sales is to attract buyers. The average home buyer visits only seven builders before making a purchase decision.*[1] *Thus the business challenge for builders is to maximize the number of home buyers in their market that include their company on a short list for consideration. Once buyers are attracted, the next goal is to get them to buy a home. Effective sales skills convey a relevant and compelling value proposition. The final goal is to keep customers for life. Surveys show that the cost of acquiring new customers runs 8 to 10 times higher than the cost of keeping existing ones.*[2]

How. *Tried-and-true techniques mark the path to achieving your sales goals. First, effective marketing requires the right messages and strategies to attract the most customers. This is a rapidly evolving art or science (take your pick) as new social media and marketing tools continue to emerge. Once buyers are attracted to a builder's product, point-of-sale interaction makes it personal and relevant to each customer. Sales skills are critical, but effective handout information and on-site displays can assist this personal interaction. Homes are such a large purchase, effective sales must incorporate appropriate financing solutions to complete the transaction. The process does not end with the sale. Customer services are critical to ensure maximum satisfaction that can keep each buyer a customer for life.*

Builders who apply the retooling concepts in this book will be taking housing to a substantially new level of excellence. For the first time, homes will be comprehensively better, not just bigger with cosmetic touches. Community development, design, performance, and construction will be combined in a unique value proposition home buyers have never seen before. The sales process has to be retooled to fully integrate the benefits of these improved homes. This will be particularly challenging because the industry sales infrastructure is familiar with selling highly visible aesthetic features such as architecture, granite counters, and large master suites, whereas most improvements in high-performance homes are invisible.

---

[1] Gord Cooke, "Houses That Work," Excellence in Building Conference, Energy and Environmental Building Association, Denver, CO, 2009.

[2] Joanna L. Krotz, "Keep Your Customers Happy and Coming Back for More," Microsoft Business for Small & Midsize Companies, http://www.microsoft.com/business/en-us/resources/marketing/customer-service-acquisition/keep-your-customers-happy-and-coming-back-for-more.aspx#Keepyourcustomershappyandcomingbackformore

What Is Effective Home Sales?

Effective home sales is a process that conveys relevant, personal value to each home buyer and then boldly, but discreetly, asks for his or her business. The key principles presented here are drawn from time-proven principles and personal observations.

Effective Home Sales Starts with Corporate Culture

I was about to face the biggest career decision of my life—whether to join the EPA and direct ENERGY STAR for new homes. With a formal offer waiting for final processing, I was invited to attend a three-day ENERGY STAR retreat at a conference center in West Virginia. I know it's an overused expression, but I definitely wasn't in Kansas anymore. I was in a geographically and cultural alternate universe. from my position at that time managing energy programs at the California Energy Commission (CEC), The CEC has many impressive accomplishments, including approving new power plants, developing state energy policy, and promulgating energy codes and appliance standards. However, the organizational culture that worked well for policy analysis, promulgation of regulations, and code implementation (e.g., exercising extreme caution, ensuring there were duplicative checks and balances, and being willing to delay all decisions until exhaustive research was completed) was poison to the needs of a successful marketing program (e.g., moving at the speed of business, empowering staff to make routine program decisions, and being willing to accept mistakes—lots of mistakes).

I found myself alone in a cabin in the middle of rural West Virginia pondering what I would be getting myself into if I were to accept this position. Change is always a challenging step into the unknown, and our family was very happy in California. I had already turned 40, was feeling disconnected from the energetic and passionate EPA staff that skewed very young, and knew opportunities for major career moves were rare. So I began making a list of positive and negative outcomes associated with taking the job. Very quickly the number of negative items began to overwhelm the positive ones.

As the conference began, one of the senior staff gave a kick-off presentation completely devoted to selling—and everything changed for me. The content was a revelation with insights that would benefit any public or private enterprise, and particularly ENERGY STAR's daunting task to convince American homeowners, businesses, and government organizations to voluntarily increase energy efficiency. It was clear this organization understood that its mission could only be accomplished with cutting-edge sales and marketing solutions and a commitment from the entire staff to embrace the fact they were "sales people." I went back to my room, tore up my list of possible outcomes, and accepted the formal job offer that came the next day. Sixteen years later, I still periodically review my notes from the session that inspired me to take the most fulfilling position of my life (see box). These ideas continue to be "words to live by" and are broadly relevant to any organization that needs to sell products or ideas . . . in other words, every organization.

EPA Internal Presentation on Sales

Selling Green Programs: We are Sales People

Presentation by Steve Andersen
September 1994, Coolfont, West Virginia

Goals of a Good Sale:
- Fair price
- Good weight (don't short-change value)
- No hidden costs
- Money-back guarantee if unsatisfied for ***their*** own reason
- Keep them coming back for more

What We Sell:
- Green products
- Global change division team
- Good government
- Reputation you can bank

Preparing Product for Sale:
- Product development
- Product packaging (different packages for different audiences)
- Sales materials
- Product use support

Empower the Client/Partner to Buy Products That Protect the World:
- Belief system
- Briefing tools and skills
- Rewards and reinforcement

Making the Sale:
- Cultivate the client
- Understand and overcome reluctance
- Make the pitch (what I have to do to convince you to buy this product—confront customer)
- Close the deal
  - What action they take
  - What agreement made
  - How know action taken

Follow Up:
- Deliver the product
- Support the product
- Reinforce the purchase (notify client of benefits and measure results, useful for testimonials)
- Improve the product

Use the Client:
- As a reference
- For presentations
- As an author (write articles with them, publish under their own name)

Sales People Attributes:
- Ambitious (work hard)
- Product passion
- Science and technology
- Sales skills

Steve Andersen's Advice:
- Corporate/cultural sensitivity complicated (Most people in green business never aspired to corporate life. They need to think about ***their*** goals, what ***they're*** doing, and not be judgmental.)
- Need optimism and compassion
- No fault, no sale (People have the right to say ***no,*** even if you are right. You want to look down the road and not burn bridges; you'll be back. Stay in touch over a long period of time.)

Immediately after that retreat in West Virginia, I obsessively committed myself to become a better salesman. This included getting immersed in the best books, conferences, and training I could find on all topics related to sales. It was fulfilling to realize that many of the natural instincts I used to sell architectural services in years past were consistent with the best sales practices advocated by experts. It was more fulfilling to address critically important gaps in my sales skills that could help make me better.

Effective Home Sales Recognizes Many Home Buyer Concerns

A major reason homes are such an overwhelming consumer purchasing decision is the broad range of legitimate concerns that have to be addressed in the sales process. Figure 6.1 maps these concerns.[3]

First, consumers care about the *company* they are entrusting with such a large and important purchase decision. *What is the company's history?* Is this an established company? How good is their reputation? Does their experience match the buyer's needs? *How good is their service?* Are they helpful in making selections? Do they accommodate special needs? Can they help arrange financing? Do they keep buyers apprised of progress? Do they stay on schedule? And after completion, do they explain how to use the home and respond promptly to concerns? *What are their core values?* Are they committed to the community? Are they committed to quality work? Are they committed to special values that the buyer cares about (e.g., sustainable construction, energy efficiency, diverse communities)?

Then come concerns about the *community*. When purchasing a home from a production builder, the buyer is typically entering a new subdivision. The community could be a new subdivision or part of an established neighborhood when purchasing a spec home from a custom builder. This concern is addressed before a home is sold when home buyers prepurchase vacant lots or "tear down" an existing home in order to construct an improved or larger new home. All aspects of the community are critical at some point in the sales process. *How good is the location?* Is there good proximity to work, preferred recreational activities, shopping, medical services, and other areas of interest? Is the neighborhood safe? Is the neighborhood quiet? *How desirable are the amenities?* Does the community include desired recreational facilities (e.g., swimming pool, golf courses, parks, sports courts, trails)? Is there a gated entry? Are there well-maintained landscaped spaces with mature trees? Is there a neighborhood town hall? *And how good are the schools?* What is their ranking? How good are their test scores? How safe are campuses? How good are the special programs? For many buyers, access to quality schools is a deal maker or breaker.

Obviously, the home itself is a major concern. Yet it is surprising how little consumers understand about evaluating their home purchase.[4] Increasingly home buyers do have a wide range of concerns. *How attractive and functional are the design and layout?* Do the architecture and floor plan

[3] Cooke, "Houses That Work."

[4] Visit my website at http://www.samshomebuyingtips.com for a guide with 10 checklists for home buyers going through the home selection process.

Figure 6.1 Home Buyer Concerns.

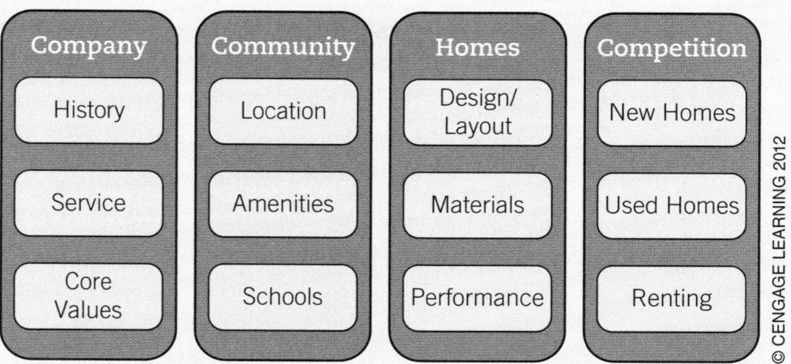

Source: Gord Cooke, Building Knowledge Canada, Inc.

meet or exceed expectations? Are there enough bedrooms? Will furniture fit in key rooms? Is there adequate daylight? Is there enough storage? Is there a garage, and what number of cars can it fit? *What is the quality of the materials?* How durable are the siding, roofing, and windows? How much trim is provided? What quality cabinets are provided? What is the quality of the appliances, lighting, and fixtures? What is the quality of the finishes? *Although consumers are often unaware of the concept, they are becoming increasingly concerned about home performance.* Is the home energy efficient? Will it be comfortable and quiet? Will this home provide a healthy environment for the family? Is the home durably built so the buyer can expect low maintenance costs? Will this home stand up to severe weather conditions, natural events, and pests relevant to this region?

Finally, consumers worry whether they made the right choice relative to other homeownership options. How does this home compare to other new homes? How do the architecture and layouts compare to other builders? Is the quality of construction better or worse? Is the home under consideration more or less efficient, comfortable, durable, and healthy than other options? How much do the various options cost to purchase and own? Do they employ the latest advances in construction technologies and practices? *How does a new home compare to buying a "used" home?* How do the benefits of used homes (e.g., closer proximity to employment centers, established neighborhoods, and urban amenities) compare to those for new construction (e.g., all new finishes and equipment, greater energy efficiency, more square footage)? *And why should the consumer invest in homeownership rather than renting?* Why should the buyer be obligated to provide maintenance and upkeep? Why should the buyer be tied down to a property and not be geographically flexible to pursue other job opportunities? Should the consumer invest in real estate when many experts don't forecast significant appreciation?

It is within this comprehensive spectrum of home buyer concerns that the retooled housing industry must sell more livable communities, better designed homes, improved home performance, and quality construction practices and technologies. This requires highly skilled sales agents who can sell what home buyers cannot see.

Effective Home Sales Addresses the Key Rules of Sales

Sales and marketing are complementary business activities that often go hand in hand, but entail very different skill sets. Marketing gets customers interested in your product or service; sales gets customers to buy your product or service. Marketing by design cannot be personal. It is a broad outreach effort to the universe of potential customers that might be interested in your products or services. The right marketing message will convey the best generic value proposition for a diverse home buying audience. Selling, on the other hand, is extremely personal. It is all about understanding and addressing each individual customer's needs and wants. Marketing delivers the prospects; selling closes the deals. This section focuses on 10 key rules of the sales process. They are derived from my experience selling architectural services, extensive reading on all sales-related topics, attending a wide array of sales seminars, and consulting extensively with sales experts. Many of these rules are so old they probably began with the sale of the first wheel, but they are all words to live by.

Rule No. 1: Only Two Things Are for Sale: Solutions and Opportunities. When it comes to any sales transaction, there are only two basic reasons for a customer to buy a product or service: it provides a solution to a real problem or it satisfies a meaningful opportunity directly relevant to that specific buyer. This is the same whether you are selling paper clips or new homes. In the case of new homes, buyers have an impressive array of problems that need solving: affordability, more living space, more storage, healthier living conditions, better schools, safer neighborhood, shorter commute, and the list goes on. Similarly, home buyers are pursuing many opportunities: preferred design, more space, luxurious kitchen, quality finishes, greener living, higher ceilings, and other aspirations. The challenge in sales is to uncover these problems and opportunities for each individual sales prospect and to effectively address them so the offer is personal and relevant. Retooling to more livable communities, superior quality designs, comprehensive home-performance improvements, and advanced construction technologies and practices substantially enhances the sales agent's ability to solve problems and provide opportunities.

Rule No. 2: Knowledge Breeds Enthusiasm and Earns Trust. When my wife and I were undergoing a recent kitchen remodel and it was time to shop for new appliances, we started at an appliance outlet that did not offer the lowest prices but that provided an excellent selection. One of the young sales staff introduced himself and asked if he could offer assistance. His youth was off-putting, but only for a very short time as he gently, but effectively, took control of the sales process saying, "Please follow me, I'd love to help you select a new dishwasher that best meets your needs." Of course we followed him. He immediately asked questions that helped him understand the performance and design features we considered most important (see Rule No. 4). Armed with this information, he demonstrated important technical differences between brands and models that allowed us to confidently address our purchase criteria. His impressive knowledge and enthusiasm were contagious, and we bought an entire appliance package from him. His passionate knowledge earned our trust and made any concerns about getting the absolute bottom price a secondary consideration (see Rule No. 7).

There are no shortcuts. Whoever represents a home builder at the point of sale, whether it is the builder, a real estate agent, or an in-house sales person, has to invest in the training and practice required to be completely facile with all the detailed technical improvements in the home, how they work, and why they are better. It is no longer adequate to just point out the granite countertops. Moreover, consumers are starved for good information and become excited when they feel they are getting good information from someone who is clearly knowledgeable. This can remove some of the stress involved in making any product purchase. Alternatively, when the customer knows more than the salesperson, things usually don't go well. This is a breach of trust. Today's home buyers have access to extensive information via Internet sites, and this will continually up the ante on knowledge requirements for home sales professionals.

Rule No. 3: A Confused Consumer Never Closes. In the process of earning trust, it is critical to keep sales messages simple. If the potential buyer is confused, it is easy to put off the purchase. Among many personal experiences with consumer confusion, I often think of my first attempt to purchase a digital camera shortly after they were introduced to the market in the mid-1990s. I went to a few big box electronics stores, started to shop, and then the confusion began. How many pixels should I get? How do I judge lens quality? How much optical versus digital zoom did I need, and what was the difference between them? How many standard settings were useful? What was the best battery technology? My goodness, I was just buying a camera, and I felt completely overwhelmed. As a result, I found myself simply walking away without making a purchase. It took more than a year before I bought my first digital camera. It shouldn't have taken this long, but the sales staff I encountered were not able to effectively communicate the right combination of features and price for my needs without confusing me with too many complex details and technical concepts. Confusion paralyzes the sales process.

Livable communities, good design, high-performance improvements, and quality construction technologies and practices are recommended for retooling the housing industry. When sales agents explain these improvements, it is critical that they use terms and descriptions that do not confuse buyers.

Rule No. 4: You Have to Listen to Be Heard, So Ask Questions. Many sales agents engaging customers are so excited to tell what they know, they wind up hogging the airtime as they "core dump" a generic sales pitch. This is in direct conflict with the 90/10 sales rule: listen 90 percent of the time and talk 10 percent. Consumers glaze over and lose interest when bombarded with generic details not related to "what's in it for me." Listening entails asking short insightful questions and letting the prospective customer talk. Research suggests that only six or seven questions can be asked in a typical sales interaction before the customer loses interest.[5] The goal is to elicit information that helps identify at least three critical concerns that can be integrated into a targeted sales process (see Rule No. 10). Therefore, it is important to ask the right questions that most effectively uncover solutions and opportunities personal and relevant to each customer (see Rule No. 5).

[5] Cooke, "Houses That Work."

Rule No. 5: People Buy Benefits, Not Features. It should be no surprise that consumers care about "what's in it for them" rather than generic product information. Many sales professionals neglect this basic concept because it is so much easier to just list the features. This reality particularly rang true while shopping with my daughter for a new phone. The salesperson spent most of her time quickly listing an amazing litany of features that came with each model. In fact, the salesperson was not able to target benefits that were personal and relevant to my daughter because she didn't first ask questions that identified her needs and wants (Rule No. 4). In the end, my daughter bought a phone that did not have a "SIM card" that would allow her to retain the directory of friends on her existing phone. Who knew there was a SIM card issue? The salesperson should have known. As a result, my daughter returned the phone. The critical benefit my daughter cared about was completely missed in the sales process.

The retooled housing industry product will effectively add an impressive array of features to the biggest purchase of a lifetime. However, it is incumbent on sales professionals to effectively translate these features into benefits:

- ***More livable communities*** with open spaces for children to play safely, community and architectural designs that encourage neighbors to meet, and lush landscaping that will protect future resale value
- ***Good housing design*** that results in optimum views, open spacious look and feel, low maintenance, less clutter, more natural light, plenty of storage, and weather-protected outdoor spaces
- ***High-performance homes*** that save owners thousands of dollars on utility bills; deliver perceptibly better room-by-room comfort; assure fresh, filtered air and only include safe products that can enhance healthy living; and provide peace of mind with less costly maintenance and protection from regional disaster risks
- ***Quality construction*** with visibly better quality materials, details, and trim, and assurance that only the best crews worked on your home because their work was tested and inspected

Rule No. 6: People Buy on Emotion and Justify with Facts. Think back on all your major purchases involving a salesperson, and the concept of buying on emotion should ring true. One of my favorites is the time my wife and I purchased our first upscale car in 1984. I had had a good architectural year and had accrued $12,000 cash to buy a new car (equal to more than $25,000 today). My wife and I were also feeling patriotic and decided to focus on American cars. We shopped one U.S. car dealership after another, but none of the American-made products were connecting with us emotionally. What is more interesting is that none of the salespeople hanging around outside the showroom approached us. Why was that?

When we looked across the street at a Volvo dealership, we were both struck by the new design for the 240 sedan. Finally, the incredibly boxy Volvo had gotten smooth edges that transformed the car into a much more elegant European appearance, at least to us. We ran across the street and walked around a few cars, and our emotion was oozing. Meanwhile out of the corner of my eye I saw Bud, the Volvo salesman. He must have weighed nearly 300 pounds and was doing a 50-yard dash in 5 seconds to be the first salesperson in the lot to lock us in as his customers. Why was that?

Once Bud got to us, he asked some questions and then began showing us the impressive quality features of the 240 sedan. He had a demonstration vehicle nearby and removed a door panel to reveal the rigorous steel reinforcing for added passenger protection. He then showed us that the odometer went to hundreds of thousands of miles, unlike American car odometers that turned back to zero after 99,999 miles. The implication that longer service life was engineered into Volvos was obvious. He took us on a test drive and demonstrated how the car stopped without any swerve when he slammed on the brakes at 45 mph with no hands on the steering wheel. And when we got back to the lot, he revealed the one fact that closed the deal: the car had 42 pounds of paint. Of course I'm being facetious; who knows how many pounds of paint a car needs? But Bud knew the basic rule of sales and was feeding us just enough facts to justify our obvious emotional attachment to that car. This transaction happened more than 25 years ago, and I still clearly remember these details. Beyond being a bit crazy, why is that? The answer is that this was a profound sales experience that demonstrated the powerful combination of emotion and facts.

None of the salespeople at the American car dealerships approached us because they were professionals with finely tuned senses that could detect emotion at a thousand yards. In its absence, they knew the chance of closing a deal would be next to nil. Any salesperson that approached us would be locked into a long time commitment—showing the car, taking the test drive, and then meeting in the closing room to negotiate a sale: "What would it take to get you in this car today?"—with little chance of success. Meanwhile their colleagues would be in a position to grab a much better prospect. In contrast, severely overweight Bud risked a heart attack running at breakneck speed to be the first salesperson to sell us the Volvo 240 sedan because we were transmitting emotion like a large neon sign.

Rule No. 7: Once Value Is Understood, Price Becomes Less Important. Another note about the Volvo purchase for your consideration is that we bought that car for $17,000. Somehow our $12,000 budget became less important when we understood the value. After all, we were buying one of the largest purchases of a lifetime. And like most consumers, when emotion and facts are there, we find the resources needed to make the purchase we really want. In fact, daily empirical observations suggest consumers don't just shop for lowest cost. If they did, we would all have the cheapest cars, televisions, computers, furniture, and the list goes on. But we don't. The challenge of the sales process is to identify value "hot buttons" and then translate features into benefits that are relevant to each buyer (see Rule No. 5). Don't give up at the beginning by thinking consumers only shop for the lowest first cost.

Rule No. 8: Under Promise and Over Deliver. The last lesson from the Volvo story (I promise) may be the most important. We put all of Bud's facts to the full test over the 17 years we owned that car. One day my wife left the car in neutral with the door open in a moment of distraction. The car rolled back and the open door hit the garage door wall. The steel-reinforced Volvo door was virtually undamaged, but our poor garage wall sustained significant damage. Yes, the steel reinforcing was effective. Another day I had to slam on the brakes as a driver suddenly pulled out of a driveway. The car braked in full control. Yes, the brakes were as good as demonstrated during the test drive. I don't know if the "42 pounds of paint" is just a silly claim, but over time the car did appear to look years younger than its age. All those facts promised by salesman Bud were indeed real benefits of ownership. That was the good news. The bad news was that even though we were able to push the odometer well past the 100,000 mile turnover for American car odometers, the maintenance costs to get there far exceeded anything we had previously experienced or had expected. On one critical attribute my wife and I value in cars—reasonable cost of ownership—the Volvo substantially under delivered. We have never bought another Volvo, even though our family income grew.

More to this point is how the American automobile industry took itself to the brink of financial collapse before its recent recovery. It was extremely painful to watch these pillars of American industry take no action to improve the quality, reliability, and performance of their automobiles even with decades of empirical data demonstrating comparatively poor repair records and resale value compared to their foreign competitors. In essence, the American automobile industry simply ignored the high-performance principles advocated for the U.S. housing industry in this book. Customer loyalty is a fickle but incredibly valuable asset. Over delivering on satisfaction pays big dividends for long-term profitability.

Rule No. 9: People Retain 15 Percent of What They Hear and 90 Percent of What They Experience. Many of the features proposed for retooled homes are too complex to easily convey to consumers. Short attention spans, too little time, and the sense of being overwhelmed often get in the way. It would take 15 to 20 minutes just to explain the basics of a properly insulated assembly (see Chapter 4). If any one of the necessary components is missing, thermal defects would be clearly apparent to an infrared camera, and many of these defects would be locked in for the 100-year life of the home. However, a typical home buyer would quickly glaze over at a description of proper insulation practices.

Alternatively, I could create an experience for the buyer that would take less than 1 minute and leave a lasting impression. A few insulation companies have developed a simple demonstration tool I call the "insulation bucket," which effectively mimics a complete insulation system. It's a container the size of a large plastic paint bucket lined with a couple of inches of insulation perfectly installed around the perimeter, on the bottom, and on the lid. I would tell a prospective home buyer that all of our homes include an advanced insulation enclosure system and ask their permission to demonstrate how it works. I would pull out a panic alarm used to scare off would-be attackers and activate the alarm. After a few seconds of ear-screeching noise, I would drop the alarm in the bucket, close the lid, and allow the prospective home buyers a few seconds to observe the impressive total silence while the alarm continued to emit amazing levels of noise inside the bucket. There is no comparison between a 15- or 20-minute explanation on the technical requirements for proper insulation installation and a 1-minute insulation bucket demonstration.

Figure 6.2: One-Minute Sales Process.

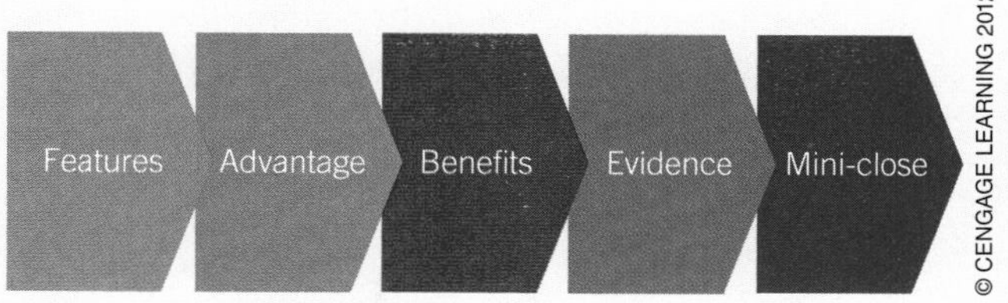

Rule No. 10: Close Often and Effectively with a Consistent Process. Sales training instructors I observe bemoan the fact that so many sales professionals don't effectively close. The concept is fairly simple. It amounts to asking for lots of little "yeses" (mini-closes) until you have a critical mass of buyer commitment and can ask for the big "yes" (final close: "Which of the remaining available lots would you like us to lock in for you today?"). The timing will be different for every buyer. Some are ready to close immediately, others may require a number of mini-closes before being ready to purchase. So how is this accomplished? The answer is with a consistent process.

When famous athletes come through and win a big game or tournament, reporters often ask how they were able to be successful under pressure. After years of listening to myriad responses, I have identified a common theme: elite athletes focus on process, not outcome. Thinking about outcome is highly disruptive because it allows the fear of failure to affect performance. Instead, successful athletes focus on the basic process they have practiced over and over and over again to ensure consistent performance. The same is true in sales. Under pressure to close a deal, a salesperson needs to stay true to a process that works. One of the best sales processes I have seen is presented by Gord Cooke. In his sales training Gord teaches a 1-minute process that consistently delivers a mini-close from the prospective home buyer (Figure 6.2).

You are not ready to begin this process until you have followed Rule No. 4 and asked questions that uncover at least three critical needs or opportunities that can be matched to features and benefits relevant to a specific buyer. For larger model home centers, it is often necessary to choreograph how to integrate these mini-sales presentations within a normal sales walk-through, including seamless links to critical displays and demonstrations (e.g., experiences).

I'll demonstrate how this process works with a complete insulation system using the "insulation bucket" experience discussed earlier. First highlight a *feature* with a strong matched benefit to that buyer and describe it in simple terms:

> *Our homes include an advanced insulation system.*

Then state a generic *advantage* why it is better, again in nontechnical terms:

> *An advanced insulation system means your home is blanketed with superior comfort and quiet while enjoying surprising low utility bills.*

The third step makes this feature relevant to that specific buyer by citing *benefits* that address specific problems or opportunities uncovered with prior questions:

> *What this means to you Mr. and Mrs. Home Buyer, is that the constant winter chill and high utility bills that frustrate you with your current home will be a distant memory.*

Then comes *evidence* that proves the benefit is real. Remember, it is critical to use experiences because they are more effective than words when conveying value:

> *I'd like to show you a sample insulation assembly that mimics how our advanced insulation system works in every one of our new homes.* [Demonstrate insulation bucket with panic alarm] *In addition, here are actual utility bills on the last 100 homes we've sold that are all hundreds of dollars below what you are currently paying each year in your much smaller home.*

Finally, always end with the *mini-close* by asking for agreement that the benefit is indeed important to that buyer:

> *Wouldn't you agree Mr. and Mrs. Home Buyer that this kind of advanced blanket of thermal and acoustic comfort should be part of every new home, particularly when it is cost prohibitive to add after a home is built?*

This process is easily delivered in one minute or less and should be fully scripted for each feature so it can be consistently delivered by sales agents. Most important, it does not take attention away from any of the core emotional features (e.g., granite counters and large master suites) because it can be inserted in the discussion while going from one room to another. Instead of wasting transition time, the salesperson accumulates critical "yeses" that this specific home delivers important value relevant to that buyer. Where prospective buyers are emotionally connected to a home, three to five key mini-closes should provide enough critical mass to move to the final close. If prospective buyers are not emotionally connected to a home, no number of mini-closes will be enough. Buyers have to be emotionally vested for this process to work (see Rule No. 6).

How Home Sales Got Here

Home Sales Have Gone from Practical to Emotional

Early American home sales were based on systems of land ownership brought over by the British. By the time the twentieth century arrived, mortgages were readily available to facilitate the process. However, getting a mortgage at that time proved to be difficult for many as the down payment required was a whopping 50 percent for a mortgage with a five-year term.[6] When the mortgage market collapsed during the Great Depression, home sales were

[6] "History of Home Mortgages," Home Mortgage Pal.com, December 28, 2010, http://homemortgagepal.com/home-mortgage-articles/history-of-home-mortgages/

limited to buyers who had saved enough money for half or all of the cost of a new home. As a result, the sales process was very practical and was based on offering buyers what they could truly afford and absolutely needed.

Under President Franklin Roosevelt, the Federal Housing Administration (FHA) was created in 1934 and quickly led to 30-year mortgages at fixed rates. However, home sales didn't substantially grow until after World War II. The combination of the GI bill, the home mortgage program, and the postwar economic boom led an explosive growth of suburban development across the country (see Chapter 2). Sales agents responded by focusing on creating emotion associated with owning your own home with a yard, more space, and the latest appliances. This was the beginning of a sales process that has been based on selling size and design trends ever since that time.

Large Builders Do Not Offer a Consistent Product

Large national production builders began to emerge in the 1960s, but regional divisions were typically given substantial autonomy on design, construction, and sales preferences. Thus these large production builders were able to negotiate economy-of-scale price reductions on products and equipment, much like a large buying co-op, but this decentralized business model made it impossible to promote companywide attributes regarding design, performance, and construction quality. The result is that large national builders to this day do not offer a consistent product across all of their divisions.

Large Builders Perfect Selling Size and Design Trends

As suburban development continued, available land was farther removed from urban centers. Because most of the profit in the housing industry comes from land development (see Chapter 2), large production builders were interested in "flipping" land as quickly as possible. The industry quickly understood that the most effective path toward this goal was to ratchet-up the emotion-driven process so buyers would accept longer commutes and the growing pains of newly developed communities. The easy emotion to tap was to increase house sizes and feature new design trends. This led to the average house size more than doubling from 1950 to 2000 (see Figure 3.13). There was constant pressure for kitchens, bathrooms, and master suites to become more spacious and luxurious. The new home footprint also increased to accommodate our growing love affair with cars and our desire to protect them. This led to a transition from street parking, to carports, to one-car garages, to two-car garages. In many upscale developments, three- and four-car garages have become standard.

Design changes effectively tapped into perceived consumer interest in more complex architecture and the latest finishes. Over time, cabinets have transitioned from stained oak, to white enamel, to stained maple. Similarly, kitchen appliance finishes have transitioned from white, to avocado (which often evokes a "what were they thinking" reaction), to black glass, to white glass, to stainless steel. And counter finishes have transitioned from laminate, to ceramic tile, to fabricated monolithic tops, to monolithic stone.

Size, design, finishes, and location at the lowest price have been the features the sales process has sold for decades. This has also become the limit of what housing industry sale professionals know how to sell.

Why Effective Home Sales Is Broken

The emotion-driven sales process adopted by large production builders attracts prospective home buyers to visit a subdivision sales office and then walk through "brand new," highly decorated model homes. This method has been perfected and substantially abused during the recent housing boom years. It is a limited approach to sales that will no longer work. A soft housing market with a massive fire-sale of distressed homes, a strong interest in efficient homes, and a growing preference for urban living are challenging this sales process. These challenges demand a greater commitment to effectively marketing the transformational improvements recommended for the housing industry in this book (see Chapters 2–5). Unfortunately, research and personal experience indicate the housing industry does not have the skills needed to sell the compelling but invisible benefits associated with these retooled homes. The housing industry sales infrastructure is substantially broken.

A Strong Brand Is a Terrible Thing to Waste

Most large production builders and many smaller builders have failed to establish an effective brand beyond size, location, and price, with a few amenities occasionally thrown in for good measure (e.g., golf courses and walking trails). This is evident in the thousands of print advertisements, billboards, signage, and Web sites currently promoting new homes. This problem can be attributed to the dominant business model wherein individual divisions operate autonomously. Thus the nation's largest home builders effectively have no brand except they are big, build in a defined price range, and offer various locations. By comparison, large manufacturers in other industries commonly develop strong brand reputations around value-based attributes (e.g., cutting-edge design, superior quality, proven reliability, outstanding service, extended warranties, better performance, high customer-satisfaction rankings, quality materials and workmanship). This is a lost opportunity to establish a meaningful corporate message associating their product with positive attributes home buyers value. It has become even more critical for builders to develop a brand advantage with the challenging economic conditions facing the nation.

Housing Sales Agents Do Not Ask Questions Effectively

In a study conducted by Gord Cooke in the late 1990s, four pairs of researchers posed as married couples looking to buy a new home. These mystery shoppers toured 200 production builder developments in Canada. The results of this study and ongoing mystery shopping in both Canada and the United States show that sales agents are extremely weak at asking probing questions.[7] Moreover, I have corroborated this finding by personal observations in

[7] Gord Cooke, personal communication, November 10, 2010.

subdivisions representing thousands of homes throughout the United States. Sales becomes a guessing process when questions that effectively uncover each buyer's critical "hot points" are not asked. Housing industry sales agents need to learn to ask effective questions that will surgically uncover the specific needs and wants of each buyer.

Features Are Not Translated into Benefits

The same mystery shopping experience just cited also reveals that sales agents and real estate professionals are not trained to translate complex technical features into simple concepts that effectively address personal and relevant benefits. Explanations of benefits are too complicated and are poorly understood. Home sales professionals have to become completely facile with all the benefits associated with retooled homes and be able to explain them in easily understood terms.

The Housing Sales Process Over Promises and Under Delivers

As mentioned earlier, the production builder sales process often relies on furnishing models with the best finishes, furnishings, and maximum number of upgraded features to drive emotion as prospective customers walk through them. Unfortunately, this functions like a bait-and-switch sales process; buyers are emotionally drawn in with interior design cosmetics and upgrades far above what they are likely to realize as owners. Buyers experience an emotional letdown when they move in because their home does not have the look and feel of the model they purchased. In addition, this emotional experience often blinds buyers to other quality construction and energy-efficiency concerns that later breed dissatisfaction. For example, my cousin bought a new home from a high-end production builder, but just replaced the windows after less than ten years because of poor quality, drafts and high energy costs. Over promising and under delivering is not a winning formula for earning loyal customers and maximizing long-term success.

Builders Are Obsessed with Lowest First Cost

Can you blame builders for wanting to minimize their expenses? Their business entails substantial costs to secure designs, construction documents, and permits; purchase and develop infrastructure on raw land; construct homes; and sell them. Then there are a plethora of risks associated with setting the right price point, satisfying exterior and interior design preferences, timing construction to capture favorable financing rates and economic conditions, and avoiding expensive warranty expenses. Obviously, lowest cost minimizes risk. Wrong! Investing additional value in more livable communities, superior quality designs, quality construction, and high performance can substantially reduce risk. By building the same old product with a facelift, the industry is maximizing risk and increasing its exposure to litigation. The U.S. housing industry makes decisions about adding improvements that are not visible based on the perception that consumers won't pay more for the added value. This misconception is based on their experience selling homes without effectively applying the rules of sales. In this soft economy, builders are likely to remain paralyzed against adding cost for the retooling improvements recommended in this book. Yet it is precisely because of the soft economy and the new normal that it is time to invest in transforming the value proposition.

Experiences Are Not Effectively Integrated in the Sales Process

Many experiences can dramatically demonstrate why high-performance homes with quality construction take new housing to the next level of excellence. Beyond the limits of the spoken word, they provide visual, high-impact evidence validating claims about critical benefits. And most important, they are easily understood even by the least informed home buyers. However, these demonstrations are rarely used in the new home sales process.

The Housing Industry Does Not Invest Enough in Sales Training

The average U.S. business invests approximately 5 percent of sales staff time in training, or about two hours per week. Personal observations and discussions with other industry experts indicate that the housing industry does not come close to this commitment. As a result, sales agents are not prepared to sell the value of retooled homes. An essential part of sales training is ongoing feedback and accountability: you can't manage what you don't measure. Unfortunately, the housing industry rarely measures the effectiveness of sales agents to sell quality and performance improvements.

The Housing Industry Does Not Own Customers for Life

Right now I'm writing with a cheap pen that only cost pennies, but it has a label identifying the manufacturer. However, I don't have a clue as to who constructed my home worth hundreds of thousands of dollars. To uncover the secret builder, I would have to visit my local building department. It often appears that home builders are intentionally trying to disconnect themselves from the owners of their homes as soon as the sales transaction is completed. A much stronger business model would be to build a product with extremely high customer satisfaction and then maintain an ongoing relationship with each home buyer and ensuing new owner. Every sale should breed more sales.

How to Fix Effective Home Sales

Build a Brand

It's time for all builders, but particularly large production builders, to make a top-down commitment to develop a meaningful brand. The following important attributes could be associated with builders who employ the retooling recommendations in this book:

- ***Sustainable communities*** that are pedestrian friendly, designed for more social interaction, enhanced with lush landscaping and open spaces, appointed with superior quality details (e.g., pavement, benches, signage, and lighting), and include several special features (e.g., vistas, parks, or recreational facilities)

- ***Good home design*** that employs regionally appropriate architectural styles including low-maintenance materials; free natural comfort that feels better; clutter-free design with generous storage closets and built-in cabinets; efficient use of space that lives "big"; advanced lighting design that enhances ambiance, control, and efficiency; high-end appliances with longer warranties; and quality fit, finish, and trim
- ***High-performance homes*** that deliver superior energy efficiency with substantially lower utility bills; healthier indoor environments that reduce the risk of mold, dangerous chemicals, radon, and combustion gases; improved durability with comprehensive water protection and longer warranted equipment; and greater disaster resistance with lower insurance premiums and better protection
- ***Quality construction*** utilizing advanced technologies, materials, and installation practices along with rigorous quality assurance systems holding construction work accountable to higher standards

Once key attributes are identified, they must be fully integrated in the marketing and sales process with effective messages and delivery strategies. This is a rapidly moving target as conventional advertising techniques used by the housing industry (e.g., signage, billboards, print, radio, and television) have to be balanced or replaced with new tools (e.g., Web sites, search engine advertisements, and social media). All of this effort will require a new level of discipline to consistently reinforce the brand message. Consumers have to know how good the retooled home has become, and this begins with effective branding.

Train Sales Agents and Hold Them Accountable

One of my favorite builders in the country is Vern McKown, president of Ideal Homes in Norman, Oklahoma, and the largest builder in the state. Ideal Homes has a longstanding commitment to high-performance homes, and for many years I've used a quote that I attribute to Vern McKown: "If you don't tell your story, you give it away." It's a simple concept, but it is amazing how many builders fail to apply it. Vern does not. He recognized his sales agents did not know how to sell the hidden value of his high-performance homes and took action. He hired expert sales trainers for ongoing training and practice sessions and balanced that with internal role-playing exercises at weekly sales meetings. In addition, mystery shopper consultants were hired to secretly videotape his agents during the sales process to ensure accountability and to assist in continual improvement.

All builders need to make similar investments in training their sales agents to ask questions, become facile presenting benefits that are personal and relevant to each buyer, and close effectively. The industry has to have the discipline to require agents to practice and to hold them accountable for applying these sales skills. It will yield impressive returns.

Integrate Experiences That Convey Value in the Sales Process

As discussed earlier, experiences are a powerful sales tool that is substantially neglected. This is especially true for design, home performance, and quality construction improvements that are not visibly obvious to the home buyer. Builders need to explore all opportunities to add experiences in the sales process, including working with manufacturers who have developed displays that demonstrate the effectiveness of their products. Table 6.1 lists some experiences that can be used to sell comprehensive building science. Note that a consumer friendly sales term is shown for each building science measure listed. Words matter!

Sell Value

In the early 1970s, American cars were sold by visiting a dealer, selecting a car, and then being forced to make a large number of painful choices about which features you could afford from among the many features you wanted. Optional features were often tacked onto the American car rather than being fully integrated in the interior design. Japanese manufacturers, particularly Honda and Toyota, introduced a complete paradigm shift in this sales model that helped initiate their surge in market share. Their cars came loaded at the base price with the most desired and popular features, and these features were fully integrated in the interior design for a superior aesthetic look and feel. The entire car purchase experience felt completely different and was more satisfying.

The U.S. housing industry needs to adopt a similar paradigm shift in their sales process. Builders should incorporate high-quality finishes and features into the base price of each home. Thus builders will be selling homes that won't disappoint their new owners on move-in day. This is a much stronger business model for long-term growth. It's still appropriate to fully decorate model homes so they look their best and to offer luxurious upgrades, but the quality of standard design features should never be compromised to sell high-profit upgrades. What a concept; all homes sold by a builder should reflect well on the company. This will be a difficult pill to swallow for an industry addicted to the short-term profit provided by high mark-up upgrades.

Provide a Badge of Honor

New homes and communities that follow the retooling recommendations are better for the country by providing cleaner air for all Americans (less energy consumption means less pollution burning fossil fuels), improved national security (less reliance on foreign energy sources), and increased jobs (jobs building more efficient homes cannot be sent overseas). Much the way special insignias and often unique designs on hybrid electric cars function as a badge of honor for environmentally conscious car buyers, high-performance homes that are right-sized and part of more sustainable communities should be visually marked with a

(Continued on page 204)

Table 6.1 Sales Experiences for Building Science Measures

BUILDING SCIENCE MEASURE	SALES NAME	EXPERIENCE TECHNIQUES
Air-tight construction	Eliminate holes	Show an open window and explain that the typical home has cracks, holes, and penetrations that add up to a huge gaping hole. Ask prospective buyer to imagine all the bugs, pollen, dust, humidity, and cold/hot air that can continually enter through an opening this big. Then close the window to a small crack and explain that this home has advanced air sealing materials and practices that effectively close the huge hole to a very small remaining opening. Gather comparative infrared images that effectively demonstrate thermal defects with leaky doors, windows, sill plates, penetrations, and incomplete air barriers in "used" or minimum code homes compared to high-performance home images that are thermal defect free.
Air-tight ducts	Healthier "lungs" of your home	Set up a display with two duct assemblies at least 10 ft in length. One should be constructed with leakage found in typical "used" homes and the other air sealed to high-performance standards. Theatrical smoke can be injected into both ducts with a fog of smoke pouring out of the leaky duct while the tight duct has no smoke leaking. It will be an emotional experience to see the wasteful and expensive loss of heating and cooling visually demonstrated. Simple questions should be asked during the demonstration such as "How big a hole would you like in your duct system wasting money you paid for heating and cooling?" After the typical response, "no holes," you should respond, "Wouldn't you agree that all homes should be built with ducts this tight?" Gather comparative infrared images that effectively demonstrate thermal defects with leaky ducts in used or minimum code homes compared to tight ducts in high-performance homes.
Complete insulation system	Insulation that works	Place an extremely noisy panic alarm inside an insulated bucket that mimics a complete insulation system, close the lid, and demonstrate the complete silence delivered with an advanced insulation system. Where using spray foam insulation, provide a manufacturers' display showing three different insulation assemblies equipped with fans and lightbulbs that shows Ping-Pong balls float over fibrous insulation because air flows through it, but the Ping-Pong balls stay put over foam because air cannot flow through it. A digital thermometer inside the insulation also shows that foam insulation is much cooler than fibrous insulation in the absence of an air barrier and thus much more effective at resisting heat flow from the lightbulb. Gather comparative infrared images that effectively demonstrate no thermal defects with the complete insulation system compared to egregious defects in used homes where insulation is installed with gaps, voids, compression, and misalignment.
		(Continued)

Table 6.1 Sales Experiences for Building Science Measures (Continued)

Building Science Measure	Sales Name	Experience Techniques
Low-e windows	Advanced windows	The easiest experience is to have a home buyer simply touch a low-e window exposed to direct sunlight to feel how relatively cool it is to the touch. This demonstrates how advanced window technology invisibly provides better comfort. Use a high-performance glass manufacturers' display with identical lightbulbs behind clear glass and low-e glass. Radiometers with fan blades that rotate in proportion to incident heat are in front of both glass panes. When the lights are turned on, the radiometer in front of the clear glass is spinning so fast the blades are not visible, and the one in front of the low-e glass is spinning so slowly that it looks like slow motion. This demonstration easily translates a complex technology into improved comfort and low bills by simply saying, "Think of those blades as the spinning wheel on your electric meter and imagine all the money you will save." Low-e glass blocks 90 percent of ultraviolet light that is very damaging to window coverings, furnishings, and finishes. Provide a display that shows two identical pieces of fabric that have been left in front of clear glass and low-e glass for two years. The one in front of clear glass is completely faded, and the one in front of low-e glass has near-full color. This demonstrates at a glance the impressive durability benefits with low-e glass.
HVAC quality installation	Engineered comfort system	Provide copies of the HVAC Contractor Checklist and HERS Rater Checklists used for ENERGY STAR Qualified Homes depicting comprehensive measurements and tests performed to ensure each system is properly sized, components matched, adequate air flow is delivered to all rooms, refrigeration is properly charged, and the duct system is properly installed. "Shouldn't all new homes have their central comfort system quality assured to this level?" Before walking into a bedroom, show the pressure balancing measure (transfer grill, cross-over duct, or dedicated return) that allows air to flow out of the bedroom to the central return, and explain that most homes do not have this important feature. Show how much harder it is to blow up a balloon as it is increasingly filled with air. Explain that bedrooms can experience comfort and moisture problems without pressure balancing that lets air flow out of the room when doors are closed.
Whole-house ventilation	Fresh air system	Show pictures of dirty and dusty attics, crawl spaces, and garages. Then explain that too much air with dust, pollen, pests, and moisture can come from these types of places without a fresh air system. "Wouldn't you agree it's important to have controlled and filtered fresh air coming into your home?"

Table 6.1 Sales Experiences for Building Science Measures (Continued)

Building Science Measure	Sales Name	Experience Techniques
Indoor air quality	Healthier home	Provide a "nutrition label" with your home that lists all the radon, formaldehyde, volatile organic compounds (VOCs), molds, moisture, pests, and carbon monoxide not included in the home because of comprehensive indoor air quality improvements compared to the average levels of these pollutants found in typical homes. Explain that no lead or asbestos is used in new construction, but these materials are commonly found in homes built before 1978. In fact, any home improvements on homes of this vintage are likely to require expensive lead abatement measures per new EPA regulations. This demonstrates that the home can be healthier and calls attention to significant air quality issues associated with used and other new minimum code homes. Create graphic images with stacked 42-gallon oil tanks, gallon containers, or paint buckets to depict the large volumes of carcinogenic formaldehyde and VOCs included in a typical home that have been eliminated in your homes. Show the mechanical ventilation system with a sign next to it indicating it delivers over 100,000 cubic feet per day of fresh, filtered air (actual number varies by home). Show a giant image of dust mites (they are grossly ugly) with a caption that billions of these pests appear in many new and used homes, causing serious respiratory problems. However, dust mites are prevented with comprehensive moisture control and water protection systems found in your homes. You may want to cite the fact that one-third of 14 million doctor visits per year are associated with dust mites.
Water management system	Water protection system	Show images of dry rot, mold, and other moisture damage that have occurred at windows and doors without pan flashing, roofs without heavy membranes at valleys and eaves, walls without weather resistant barriers, foundations without capillary breaks, and drain tile without fabric filter wrapping. Explain that it is common practice in older homes and many new homes to leave out many of these critical water protection details. Images are available from the DOE Building America Web site and other sources of building science information. Use demonstration with different house-wrap material sealed to the top of glasses full of water. Puncture the different materials with a nail while mentioning thousands of nails are used to adhere house-wrap to a home. Then tip the glasses upside-down and show the glass with the quality house-wrap is leak free and the off-brands have a constant drip. Home buyers will quickly understand the
		(Continued)

Table 6.1 Sales Experiences for Building Science Measures (Concluded)

Building Science Measure	Sales Name	Experience Techniques
Water management system (continued)	Water protection system (continued)	benefit of a better quality weather resistant barrier. Then explain that your company applies the same quality selection process to a full system of water protection details. "The National Association of Home Builders research center reports that water damage is the most common maintenance problem experienced by homeowners. Shouldn't every new home have this level of protection?"
Disaster resistant construction	Disaster protection	Tell buyers of the 10 percent discount for home insurance because the home meets "Fortified Home" criteria. This demonstrates that the insurance industry recognizes substantially reduced risk of damage for homes with rigorous disaster resistance improvements.
Third-party verification	Quality assurance	Provide rating reports and documentation of testing results that demonstrate the home was held to a higher standard of performance. Pictures of testing and inspection work demonstrate workers were held accountable to a higher standard of performance.
Overall high-performance home qualifications	Peace of mind	Back up long-term performance relative to indoor air quality, utility bills, comfort, and durability with a 30-year warranty because you can when comprehensive high-performance improvements are included, and your competition can't when they are omitted. ENERGY STAR is a government-backed label that a home meets strict guidelines for energy efficiency and includes comprehensive building science improvements. EPA's Indoor airPLUS is a government-backed label that a home includes a comprehensive package of indoor air quality improvements. EPA's WaterSense is a government-backed label that a home meets strict guidelines for indoor and outdoor water conservation improvements. Green labels (LEED for Homes, NAHB National Green Building Standard, EarthCraft, and a variety of regional programs) indicate home improvements included are compliant with a specified level of "green" performance. Fortified Home label is an insurance industry label that ensures a home has rigorous improvements to address regionally specific disaster risks.

badge of honor. This can be done with special signs or flags at the entrance to communities, house plaques mounted next to the front door, and window decals. The fact that homes and communities are special should not be hidden.

Own the Customer for Life

Substantial work and investment are required to earn each customer, so don't break the relationship when you hand over the keys to a home. Many hours and dollars were spent developing and executing marketing strategies, providing sales support for each customer, and in many cases facilitating the transaction process. Conventional builders seem to avoid long-term customer relationships, possibly in an effort to avoid service calls for defective work. This may be a reasonable concern where business-as-usual operations emphasize lowest first cost at the expense of complete building science and quality construction. However, with retooled homes the goal flips to strenuously working to keep a permanent relationship with each and every buyer. Three specific recommendations for owning a customer for life are discussed in the following sections: providing after-sale education, backing up superior homes with a superior warranty, and adding a full service center operation.

Provide Customer Education. As discussed earlier, you may be able to get three to five key features and benefits integrated into a normal sales process. However, there is so much more story to tell with retooled homes that would yield benefits in terms of referrals and customer satisfaction. The reality is that buyers are not prepared for a lot of detail during the purchase process when emotion is too distracting. After the sale, buyers have sobered up and need lots of good facts to justify the very scary decision they have made. This can help offset tendencies for "buyer's" remorse when homeowners begin to confront how they will pay for their home. In addition, there is often a sense of loss after a purchase because the excitement of buying a new home comes along very infrequently in life, and that chip was just cashed. Special care and attention is needed, and builders need to proactively intervene as soon as the purchase contract is signed.

The after-sale seminar is one effective technique for addressing home buyer education. All recent buyers, and possibly even prospective buyers from a defined period (e.g., the last one or two months), are invited to a special training session. Food or gifts (e.g., passes to the local movie theater) can be used as enticements. There are two options for selecting a speaker to present the features and benefits of the retooled home: bring in a charismatic professional speaker who is an expert in home performance and quality construction or use an in-house spokesperson. If you have an enthusiastic and extremely knowledgeable staff member who is an excellent speaker, I recommend going with that option. That person may be less professional than the "celebrity" speaker but may be in a position to speak more from the heart as a representative of your company. If a builder doesn't have a person with the necessary skills, he should use an expert speaker, but make sure the content focuses on benefits associated with his specific homes rather than generic material. With either option, this is an opportunity to deliver a more comprehensive story about the builder's radically improved new homes that will deliver dividends in terms of referrals, customer satisfaction, and reduced callbacks. This is also the perfect time to explain how to keep the homes operating at maximum performance.

Another great opportunity for educating buyers is to take them out on a tour of homes under construction. This is sometimes referred to as the "Dirty Boots Tour." It enables each builder to visually show the construction differences between retooled homes and minimum code and used homes. It is also an excellent opportunity to invite home energy raters and top subcontractors to help answer technical questions and show off advanced diagnostics and construction practices.

The final recommendation for consumer education is to provide a complete homeowner's manual. Every car comes with an owner's manual. Although no one reads it before purchase, it serves as an excellent reference that stays with the owner for the life of the car. Manuals also provide the opportunity to lay out in detail all the features, maintenance requirements, and periodic upkeep needed to maintain peak performance and often to keep the warranty in force. A home is a much bigger purchase than a car, yet it rarely comes with a full owner's manual. It's time for that to change. A superior designed home with comprehensive high-performance features and advanced construction needs to be fully explained. A manual should include how to maintain the exterior and site to ensure water protection; seasonal home maintenance that should be performed; how to operate and maintain the heating and cooling system to ensure maximum comfort; how to maintain the ventilation system so it continues to supply fresh air; how to maintain air sealing components (e.g., weather-stripping and gaskets) to ensure tight construction; how to maintain exterior finishes for maximum durability; and how to integrate future projects such as finished basements, decks, and patios.

Back Up High Performance with a 30-Year Warranty. Most home buyers have experiences living in used homes. Consequently, they bring many fears with them when making the largest purchase of a lifetime: high utility bills; feeling cold in winter; wet basements; musty smells; excessive pests; future resale value; healthy indoor environment; excessive noise; drafty rooms; cold and hot floors; damage behind walls that's not visible; the presence of mold; need for radon mitigation; and poor construction quality. High-performance homes directly address these fears, making competing used and minimum code homes obsolete. But better is not good enough. High-performance homes can become a game changer, but only if builders back up their performance.

Table 6.2 High-Performance Home 30-Year Warranty

HEALTHY AIR	AFFORDABLE COMFORT	DURABILITY
Lead free*	$60 per month average heating and cooling bills*	No moisture damage to structure*
Asbestos free*	Even room-by-room temperatures*	Dry basements and construction*
Particulates filtered to 3 microns*	No outdoor drafts*	No thermal defects*
Mold free*	Outside noise reduction*	90 percent UV sunlight blocked
Combustion gas free	No excessive humidity	No window condensation*
150,000 cf per day fresh/ filtered air*		No termite damage to structure*
VOC free*		Windows
Formaldehyde free*		Roofing
Pest free*		Siding
Radon free*		

Note: The asterisks reference specific limitations, operating conditions, and requirements for homeowner maintenance.

The big idea is to completely upgrade the new home warranty to 30 years. This will be a profound competitive advantage. For decades the housing industry typically has been offering a 1-year warranty. A significant market response should be anticipated with retooled homes that are boldly backed up with a 30-year warranty. If the average home lasts well over 100 years, this coverage represents less than one-third of a home's life. Compare this with the state-of-the-art 100,000-mile power train warranty for new cars, which is often more than half of that product's useful life. The coverage that could be included in a high-performance home warranty is shown in Table 6.2. Builders' attorneys will go ballistic, but this is really a no-risk proposition for true high-performance homes. Here's why.

No-Risk Healthy Air Warranty

Lead and asbestos are a serious risk in homes constructed prior to 1978, but they are no longer used in new construction materials. Thus this is a freebie differentiator from older used homes. Take it and make it an issue, particularly with EPA's new lead abatement laws that require certified lead abatement contractors to provide costly removal and testing services with any remodel work on older homes. Effective filtration can be warranted because High-MERV air filters remove particulates 3 microns or greater in the HVAC air stream. Mold and moisture damage can be warranted because moisture cannot occur in construction assemblies with water managed construction details. Protection from combustion gases can be warranted because direct-vent water heaters and furnaces cannot back-draft or experience flame roll-out. Large daily supplies of fresh air can be warranted because whole-house ventilation systems complying with the latest industry standard are field verified for correct air flow. Protection from dangerous chemicals can be warranted because products are specified without VOCs and formaldehyde. Limited biological contaminants can be warranted because moisture control measures ensure dust mites cannot thrive, substantially air-tight construction limits pathways for pests, and screens at vents prevent the entry of rodents. Safe radon levels can be warranted because radon gases in the soil are passively vented and can easily be augmented with power venting if necessary.

No-Risk Affordable Comfort Warranty

A low monthly heating and cooling bill warranty with plenty of buffer is easily calculated with simple load calculations (the $60 amount in Table 6.2 is a theoretical example). It can be warranted because similar low utility bill warranties have already been provided with more than 100,000 homes as part of the building science programs discussed earlier. We know these homes have consistently low utility bills. Excessive variation in room temperatures, outdoor drafts, significant noise transmission, and excessive humidity can be warranted because of the comprehensive air, thermal, and moisture flow measures included and the laws of physics that apply with those measures.

No-Risk Durability Warranty

Protection from moisture damage and wet basements can be warranted with the same comprehensive water management construction details that ensure healthier indoor air. Thermal defect-free construction can be warranted with a complete insulation system (e.g., air sealing, quantity of insulation, quality installation of insulation, complete air barriers, and minimum thermal bridging). Control of damaging sunlight can be warranted with low-e windows that block 90 percent of UV sunlight. No condensation on windows can be warranted with the combination of moisture control measures and low-e windows that are warmer in winter and cooler in summer. Protection from termite damage can be warranted where structural materials are specified that are not edible by termites. And lastly, quality windows, roofing, and siding can easily be warranted to last 30 years.

Why should high-performance home builders back up their homes? The answer is because they can, and the competition cannot. It's that simple. To prove it can be done, Toyota Homes, a prominent builder in Japan, offers a 60-year warranty on defects and performance in their homes (see Epilogue).

Complement 30-Year Warranty with a Service Center Business Model

The housing industry needs to completely rethink its business model and offer ongoing maintenance and improvement services to home buyers. I recognize that setting up a service center is a radical change, but it goes hand in hand with offering a long-term warranty. Both would be firsts in the history of U.S. housing, but it's time. Builders constructing retooled homes will want to have long-term relationships with their customers, and the 30-year warranty provides a compelling reason for this to happen with each owner. Homeowners will be required to have periodic check-ups to keep their warranty in force, and the service center is the formal relationship tool for that action. At least that is what I am proposing. This provides a powerful combination of benefits for builders.

First, it creates a consistent source of revenue that can help with the inevitable ebb and flow of the housing business. Revenue would be generated by annual check-ups along with necessary services and repairs. Check-up items would include ensuring the right filters are being used on the HVAC system, performing annual HVAC system tune-ups, verifying set-points and number of occupants meet heating and cooling bill warranty parameters, providing routine maintenance on the whole-house and spot ventilation systems, checking that all vent intakes and exhausts are clear, verifying site drainage is working, cleaning gutters, and monitoring condition of roofing, siding, and other exterior finishes. These service calls could also be used to generate additional contract work for subcontractors on projects such as finished basements, exterior and interior painting, gutter cleaning, power-washing, new decks and patios, additions, refinishing work, and additional built-in cabinetwork.

Another benefit from check-ups is that they give builders an excuse for continual communication with all home buyers. This communication can be used to inform them of new projects completed by the builder that might be relevant to friends and family, new services being offered, and important guidance for maintaining their homes so they continue to reflect well on the builder. In addition, every builder at closing should insist that buyers sign a release allowing the builder to access the buyer's utility billing data from the local utility(s). This waiver would be presented to home buyers as a tool for the builder to continually verify that their home performance meets promised standards of excellence. These data could be used to generate an annual energy-efficiency score for each home, comparing it with other like homes in the region using a free tool developed by the EPA called the ENERGY STAR Home Energy Yardstick.[8] A score can be generated for each customer using utility billing data and a few additional inputs (e.g., number of occupants, square footage, and address).

[8] The ENERGY STAR Home Energy Yardstick is available at http://www.energystar.gov

By consistently demonstrating the outstanding performance for their homeowners compared to the rest of the housing stock, builders will have a great "good news" story to accompany the reminder notice for the annual check-up. In rare cases when the energy-efficiency score is low (e.g., most other homes in the region are more efficient), builders have an important opportunity to intervene and work with the home buyer to diagnose the root cause. Low energy-efficiency scores allow the builder to come in as a "white knight" and identify a problem and provide a solution. For example, if a cable company contractor accidentally dislodged one of the attic HVAC duct connections during an installation, the resulting high energy bills would likely show up on the score. Similarly, if a homeowner purchased several large screen televisions and installed a swimming pool over the course of a year, the builder has an opportunity to explain that the high utility bills are due to the energy-intensive amenities added by the owner and are not a problem with the heating and cooling system.

It is a good idea to retain customer relationships with homeowners who will be delighted with their retooled homes. People do buy more than one home, and they can influence home purchases by other family members and friends. It's foolish to give away this asset. A service center helps keep customers for life.

It's Time to Recognize Value in the Transaction Process

Retooled homes and communities provide substantial risk reduction associated with more viable communities, superior designed homes more likely to stand the test of time, more affordability, improved indoor air quality, improved durability, greater disaster resistance, state-of-the-art technologies, and rigorous quality assurance. However, elements of the transaction process (appraisals, mortgages, and insurance) are indifferent to these retooled home improvements. These risk-based financial products all benefit from a completely different value proposition: better position for higher resale value, lower cost of ownership, lower health costs with comprehensive control of pollutants, lower maintenance costs, lower risk of damage from disasters, and higher quality construction. Thus, the entire home transaction process needs to be revamped.

Beyond the risk reduction associated with each home, these retooled homes are better for the country. They create thousands of high-paying jobs that cannot be outsourced to other countries. This includes more value-based jobs needed to design communities and homes; produce more building products; install materials and systems properly; and install disaster resistance measures. High-performance, energy-efficient homes are better for national security because they help reduce the country's dependence on foreign energy sources. This will be increasingly important as transportation competes with homes for electricity with the impending growth of electric cars. High-performance, energy-efficient homes contribute to a healthier environment by reducing the combustion of fossil fuels.

Thus zero value and risk reduction for retooled homes is the wrong assumption for appraisals, setting mortgage rates, and calculating insurance premiums. However, it would take decades for financial and insurance institutions to research and calculate actuarially based values when they need to take action now. The housing industry can't and shouldn't have to wait decades for a prudent market-based response to this retooled product. Thus it is suggested that leaders in the appraisal, lending, and insurance institutions issue top-down policies to fix this market failure. The following specific policies are recommended as placeholders.

- **The appraisal institutions should add the present value of the monthly energy savings based on accredited computer software calculations to the traditional appraised value. Fannie Mae has already set a precedent for this policy with their past energy-efficient mortgage (EEM) product. The present value calculation applies a discount rate to the future value of the monthly energy savings over a predetermined time. In the case of the Fannie Mae EEM, they used the national 30-year mortgage interest rate on January 1 as the discount rate for that year, and a 23-year time period.**
- **Lenders should provide a half-point interest rate discount for high-performance homes. This is a reasonable placeholder similar to programs offered in regional markets in recent years (e.g., Nebraska and Alaska state incentives for energy-efficient new homes).**
- **Insurance companies should offer a 10 percent discount for high-performance homes. This is similar to the discount they already offer for "Fortified Homes" meeting comprehensive disaster resistance guidelines.**

Chapter 6 Review

So What's the Story?

The explosive growth of suburban development following World War II created the need to sell mass-produced homes. The ensuing sales process appealed to many emotions: homeownership for the masses was made possible with new government policies and mortgage programs; confidence that appreciation was a "sure thing"; and the appeal of rapidly changing design trends. As a result, the new home sales process relied on dressing up models with the latest cosmetic features and finishes, with minimal scrutiny of what was behind the walls. Today the sales infrastructure is substantially limited to showing buyers visible features: "Aren't these granite counters lovely?" The industry now needs to provide a quantum leap in sales skills. The effective home sales story can be summarized as follows.

- **What It Is.** Effective home sales is a process that conveys value relevant and personal to each home buyer and then boldly, but discreetly, asks for his or her business.

- **How It Got Here.** Production builders looking to "flip" land quickly for maximum profit developed an emotionally driven sales process that shows off model homes dressed up with opulent decorating and upgrades most buyers cannot afford.
- **Why It's Broken.** The retooled housing industry will demand skills, selling value-based improvements buyers cannot see.
- **How To Fix It.** Builders must invest in developing sales skills, holding agents accountable, and backing up their product.

The details are included in Table 6.3.

Table 6.3: Effective Home Sales Summary

WHAT IT IS	HOW IT GOT HERE	WHY IT'S BROKEN	HOW TO FIX IT
Effective home sales starts with corporate culture. Effective home sales recognizes there are many home buyer concerns. Effective home sales addresses the key rules of sales: 1. There are only two things for sale; solutions and opportunities. 2. Knowledge breeds enthusiasm and earns trust. 3. A confused consumer never closes. 4. You have to listen to be heard, so ask questions. 5. People buy benefits, not features. 6. People buy on emotion and justify with facts. 7. Once value is understood, price becomes less important. 8. Under promise and over deliver. 9. People retain 15 percent of what they hear and 90 percent of what they experience. 10. Close often and effectively with a consistent process.	Home sales have gone from practical to emotional. Large builders do not offer a consistent product. Large builders perfected selling size and design features.	Builders have not developed a strong brand for their products. Housing agents do not effectively ask questions. Features are not effectively translated into benefits. The housing sales process over promises and under delivers. Builders are obsessed with first cost. Experiences are not effectively integrated into the sales process. The housing industry does not invest enough in sales training. The housing industry does not own customers for life.	Build a brand. Invest in training sales agents and holding them accountable. Integrate experiences that convey value in the sales process. Sell value rather than cosmetics. Provide a badge of honor. Own the customer for life. Provide customer education: • After-sale seminars • "Dirty Boots" tours • Owner's manual Back up high performance with a 30-year warranty. Complement 30-year warranty with a service center business model. Recognize value in the transaction process.

7

Putting It All Together

Putting this book together has been difficult because I am effectively telling an industry I love that every aspect of bringing new homes to a piece of land needs to be substantially retooled. Many may question whether these recommendations are justified when some indicators suggest the housing industry has made impressive improvements. For instance, J.D. Power's customer satisfaction surveys indicate that the number of defects in home construction have decreased significantly since the current housing slump took hold in 2006.[1] The problem is that incremental reductions in defects will not get the job done. The new normal discussed in Chapter 1 requires substantial innovation for new homes to compete with the massive fire sale of used homes and to make homeownership much more compelling. This is especially true for the next wave of Gen X and Gen Y consumers entering the home buying market.

In this chapter I summarize my recommendations and create a fictional marketing description for a hypothetical production builder community that follows those recommendations. The purpose of this chapter is to package all the "big" retooling ideas and to illustrate how they can dramatically enhance the competitive advantage for production builders who follow them.

Retooling the American Housing Industry: The Big Ideas

Some of the big ideas listed here require subtle improvements; others entail a paradigm shift. Together they represent a comprehensive change in how the housing industry does business.

[1] "J.D. Power and Associates Reports: Satisfaction with New-Home Builders and New-Home Quality Reach Historic Highs, as Home Builders Respond to Tough Market Conditions by Improving Products and Service," J.D. Power, September 15, 2010, http://businesscenter.jdpower.com/news/pressrelease.aspx?ID=2010177

© IStockphoto.com/BEANS-

© IStockphoto.com /123render.

Sustainable Land Development

Big Idea: You only have one time to make a good first impression, and this is true with housing developments as well. All the little things add up to an immediate perception of richness, natural beauty, and livability. This first impression will last for centuries because good development practices keep getting better with time. The big ideas for sustainable land development include the following:

- **Optimize the street layout for south orientation**
- **Invest in open space**
- **Invest in trees and landscaping and experts to do it right**
- **Invest in quality hardscaping**
- **Set up Home Owner Association maintenance covenants**
- **Lock in integrated quality design features with a plan book**
- **Establish special reasons to live in a development**

Good Housing Design

Big Idea: Compared to their bloated counterparts from the early 1970s, cars today are right-sized to be much smaller but still feel spacious with lots of storage, amenities, and improved quality features. It is time for mainstream new homes to follow this same path of right-sizing and then investing resources saved into higher quality trim, finishes, components, and construction. The big ideas for good housing design include the following:

- **Invest in experts**
- **Return to regionally responsive roots**
- **Right size the right way**

- **Design to a higher standard than complexity and fads**
- **Fully integrate all systems**

High-Performance Homes

Big Idea: The housing industry can deliver a better home that is more comfortable, healthier, durable, and disaster resistant at lower cost. The big ideas for high-performance homes include the following:

- **Commit to high-performance homes from the top down**
- **Invest in risk reduction**
- **Invest in comprehensive building science**
- **Own the holes**
- **Invest in energy-efficient components**
- **Invest in pollutant control**
- **Invest in disaster resistance**

Quality Home Construction

Big Idea: It is time to invest in all aspects of quality construction to realize substantially greater customer satisfaction and reduced service center costs. The big ideas for quality home construction include the following:

- **Invest in comprehensive construction documents**
- **Invest in a building science manager**
- **Invest in new technologies and practices**
- **Invest in quality assurance processes**
- **Invest in lean construction practices**

Effective Home Sales

Big Idea: The sales process has to be completely reinvented for the radically improved product possible with a retooled housing industry. Used homes and minimum code homes are obsolete; the American home buyer just doesn't know that yet. The big ideas for effective home sales include the following:

- **Build a brand**
- **Train sales agents and hold them accountable**
- **Integrate experiences that convey value in the sales process**

- **Sell value rather than illusionary cosmetics**
- **Provide a badge of honor**
- **Own the customer for life**
- **Provide customer education**
- **Back up home performance with a 30-year warranty**
- **Complement the 30-year warranty with a service center business model**
- **Recognize value in the transaction process**

Marketing High-Performance Homes to American Consumers

There are compelling benefits for buyers of homes provided by a retooled housing industry. The challenge is how to communicate these benefits simply without confusion. Of course, there are a lot of marketing tools, including brochures, Web sites, blogs, or other social media outlets through which you can get your message out. Every aspect of the homeownership experience is being taken to a new level, and buyers need to be effectively educated about what that means to them. To illustrate the potential value proposition, Figure 7.1 provides a brochure script for a hypothetical high-performance home development. The question for builders to ask is this: "Could I compete with this product?"

This marketing brochure is selling a hypothetical new home development that offers superior quality construction at lower cost with exponentially greater warranty coverage. I'll buy that—and so will most American home buyers. Used homes cannot compete with these standards. This takes the "new normal" head-on by completely changing the new home value proposition. I believe applying the recommendations in this book can completely transform the housing industry to a happier place with unprecedented customer satisfaction and dramatically lower customer service center expenses.

Change is hard. But it is time.

Figure 7.1 Sample Marketing Brochure.

Welcome to a Revolution in Housing

If you are reading this brochure, you are already familiar with the beautiful communities and homes we build. The next few pages explain critical new developments in design, quality, and performance that you can't see, but will transform the homeownership experience. As you know, homes are usually the biggest purchase of a lifetime. . . .

It's time to hold them to a higher standard.

Lifestyle Communities

If you look closely at the details,
you will understand why our communities feel and age better.

- ***Lush Landscaping***

  You owe it to yourself to visit our established communities to experience first-hand the difference expert landscape design can make as neighborhoods mature. Entries, front yards, and open areas throughout each community have been carefully planned for year-round color, diverse combinations of trees and shrubs, and compatibility with local climate to maximize durability and minimize water consumption. And since all of this landscaping is expertly maintained by your Home Ownership Association, the community is immaculately groomed year-round and from year to year.

- ***Beautiful Streets***

  Our planners are held to a higher standard to ensure rich architectural interest on every street. This is accomplished by locating garages at the rear of each property so unsightly garage doors do not dominate our streets. In addition, quality fixtures, materials, and accessories are specified throughout the community, including accent pavements, street lamps, street signs, community benches, porticos, entries, and fencing.

- ***Better Home Sites***

  Our site planners ensure home sites have maximum access to daylight and year-round comfort. This is made possible by aligning streets so houses can block undesired sun in the summer while capturing desired solar heat in the winter. As a result, most of our homes benefit from as much as 25 percent free heating and cooling and natural daylight.

- ***Neighbor-to-Neighbor Connections***

  Residents are actively engaged throughout our community because every home has a front porch, trails are easily accessible, and there are plenty of open spaces for parks, playgrounds, and sport courts. In addition, a community center with a Town Hall is available for meetings, events, and other gatherings.

- ***Safe Streets***

  Our neighborhoods are laid out for maximum protection from street traffic. This includes accent pavements to highlight crosswalks, speed bumps where needed for additional safety, and beautiful walking trails throughout the neighborhood that provide easy access to neighbors and amenities separated from automobiles.

Figure 7.1 Sample Marketing Brochure (Continued)

Design Excellence

Homes should stand the test of time
rather than follow the latest trend.

- ***Classic Architecture***

  Our homes are designed with proven details, materials, and architectural features that optimize beauty, comfort, and durability. As a result, every home experiences enhanced daylight, maximum sunlight control, weather protected porches and entries, and maximum protection from potential moisture damage. Floor plans feel generous and provide maximum convenience.

- ***Value Engineered***

  Every home has been designed to more efficiently provide all the space you need with enhanced visual interest. Open layouts and varied height ceilings make rooms feel bigger and richer. Interior rooms easily flow to outdoor spaces that are an important part of your living experience. Built-in cabinets are provided throughout the home to maximize space and optimize storage. The best interior design experts have been consulted to provide exciting color palette choices for each home. In addition, we have invested in expert lighting consultants to design each home with a rich combination of fixtures and controls that allow a variety of mood, accent, and ambient lighting.

- ***Natural Comfort***

  Our site planners maximized lots with proper orientation to natural sunlight, and our architects took advantage by designing each home to effectively capture free heat from the sun in winter and block solar heat gain in the summer. This also helps to maximize the amount of daylight in our homes. The result is year-round natural comfort along with lower utility bills.

- ***No Clutter***

  Premium quality cabinets are included throughout every home. This includes floor-to-ceiling cabinet and shelf units for family rooms, dining room hutches, desk and shelf units for bedrooms, dresser and shelf units for bedrooms, entertainment centers, and bedroom closet storage systems. There is so much effective storage, your homes can be clutter free at all times.

- ***Quality Trim, Hardware, Finishes***

  We have engineered so much wasted space, materials, and equipment out of our homes that we have been able to include premium quality trim, hardware, and finishes as standard in every home while maximizing affordability. Look closely at all of these details as you walk through our homes, and you'll appreciate the difference.

Homes That Work Better

Advanced building science applied to each home delivers optimum efficiency, comfort, health, and durability.

- ***Low Utility Bills***

  It will not be unusual to experience utility bills one-half to one-third of those in older homes. The latest energy-efficient materials and construction practices have been employed in each home. This includes certified compliance with the current ENERGY STAR® Version 3 specifications.

- ***A New Level of Comfort***

  Homes routinely miss critical details for effectively insulated and sealed construction assemblies. As a result, most people don't know how comfortable a home can be. Superior thermal performance, effectively managed surface temperatures, and highly efficient equipment are featured in every home. Brace yourself; it's like that first experience in a luxury car.

- ***Healthier Living***

  Every home is Indoor airPLUS certified, meeting EPA's strict guidelines for improved indoor air quality. Critical sources of pollution are controlled by managing moisture to minimize the risk of mold and musty smells; specifying safe materials to minimize dangerous chemicals found in sheet goods, paints, finishes, and cabinets; venting radon from below slabs; and eliminating combustion gases with direct-vent furnaces and water heaters that cannot back-draft exhaust into homes. Every home also includes fresh air supply and exhaust systems to dilute any remaining pollutants, and filters in heating and cooling systems that can remove particulates down to one micron or less. If you notice, your family gets sick much less frequently and feels better living in your new home, now you'll know why.

- ***Peace-of-Mind Durability***

  Every home has added protection from potential weather damage with comprehensive flashing and drainage details at the roof, walls, and foundation. Additional air sealing details help minimize any moisture damage inside wall assemblies. Our homes are constructed with an advanced wall system that is impervious to insects and can withstand hurricane force winds above 100 miles per hour. Every home also includes windows that block ultraviolet sunlight from damaging your furnishings, floor coverings, and window treatments. And the high-efficiency appliances and equipment that come with every home often include longer warranties and better-grade components.

Figure 7.1 Sample Marketing Brochure (Continued)

Advanced Technologies

The housing industry takes up to 25 years to adopt technical innovations; we couldn't wait that long.

More than 100 impressive new technologies improve the quality of each home we build (see our Web site for a complete list). Here are a few of those innovations:

- ***Precast Concrete Wall Systems.*** Basements don't feel like basements with twice the resistance to moisture flow and more effective insulation.
- ***Structural Insulated Panels.*** This advanced wall system provides much more energy efficiency, moisture control, strength, and dimensional stability.
- ***Spray Foam Insulation.*** This advanced insulation is used at key applications throughout each home where conventional insulation would be far less effective.
- ***High-Performance Windows.*** Each home features highly efficient, low-e windows that exceed ENERGY STAR window requirements.
- ***Cementitious Siding.*** Compared to vinyl siding, this composite material looks almost identical to wood, is much more resistant to warping, and lasts longer.
- ***Water Efficient Fixtures.*** All internal fixtures qualify as Water Sense® and are highly efficient while delivering totally satisfying water flow.
- ***Structured Pumping Systems.*** Nearly instant hot water is provided to each fixture, water heater efficiency is boosted by nearly 15 percent, and waste water down the drain is reduced by 5,000 to 7,000 gallons each year.
- ***Efficient Heating and Cooling Air Handling Systems.*** Variable speed fan motors are more efficient, make less noise, and last longer.
- ***Direct-Vent Furnace and Water Heater.*** Combustion and exhaust air are isolated so dangerous exhaust fumes cannot flow back into the home.
- ***Electric Induction Cook-top.*** This advanced electric cook-top has no emissions, cooks faster than natural gas, is safe, and is easy to clean.
- ***Solar Ready Construction.*** A solar electric system can be added in the future at thousands of dollars lower cost.
- ***Electric Vehicle Charging Station.*** Every garage is equipped with a charging station ready for the new wave of electric vehicles.

Quality Construction

Our contractors send their best crews to work on our homes because they are held accountable to a higher standard.

- ***Better Materials***

  Every home is constructed with superior quality foundation systems, moisture management details, wall assemblies, insulation, air sealants, high-performance windows, advanced heating and cooling systems, fresh air ventilation, filtration, trim, hardware, fixtures, appliances, and lighting. In addition, wood sheathing, paints, adhesives, and cabinets are specified that minimize exposure to dangerous chemicals. Only furnaces that cannot back-draft combustion products into the home are selected. All framing is treated to be termite resistant for important peace of mind. And all equipment is highly energy efficient and often with longer warranty coverage. All of these better materials are standard with every home because we have reinvested cost savings from efficient design, performance, and processes. Purchasing a home no longer entails choosing between low-quality construction materials and premium upgrades.

- ***Better Practices***

  For decades builders have opted for lowest cost construction practices because home buyers cannot tell the difference. Yet the difference is vast. Better construction practices ensure insulation that actually works; heating and cooling systems that perform at rated efficiency levels; roof, wall, and foundations that are completely protected from moisture intrusion; reduced waste that contributes to lower costs; and substantially fewer defects.

- ***Better Quality Control***

  All of our contractors go through extensive training before they are allowed to work on your home. Detailed inspections and testing ensure that their work consistently meets the highest standards for excellence in the industry. Our construction teams meet on a regular basis to continually improve every aspect of how we design and construct our homes. Because we demand such high standards, we offer an unprecedented 30-year warranty for every one of our homes. Compare our warranty to the industry standard one-year warranty.

  Homes should last at least a hundred years. And that's how we build them.

Figure 7.1 Sample Marketing Brochure (Concluded)

30-Year Warranty on Every Home*

Rigorous quality design, materials, and practices
allow us to offer unprecedented warranty coverage for every new home.

Healthy Air Warranty:

- Lead free
- Asbestos free
- Particulates filtered to 1 micron
- Combustion gas free
- 150,000 cubic feet of fresh, filtered air per day
- Volatile organic compound free
- Formaldehyde free
- Pest free
- Radon free

Efficient Comfort Warranty:

- $50 to $80 per month average heating and cooling bill
- Even room-by-room temperatures
- No outdoor drafts
- Outside noise reduction
- No excessive humidity

Durability Warranty:

- No structural defects
- Roofing
- Windows
- Siding
- No moisture damage to structure
- Dry basement
- No thermal defects
- 90% UV sunlight blocked
- No window condensation
- No termite damage to structure

The question is not, how can we offer such comprehensive coverage?
The question is, what are others leaving out that they cannot?

---

* Refer to *Terms and Conditions* for coverage details.

No Disappointments
What you see is what you get. . . .
If you love our models, you will love our homes.
We don't load our models with high mark-up upgrade options that strain your budget and leave you frustrated with compromises on move-in day. Quality design features, trim, hardware, and finishes are included in every new home. Because we engineer substantial wasted space, materials, and equipment out of our homes, we can include superior quality at no extra cost.
So move-in day is not disappointment day.

Epilogue

A Little Development in Texas

An interesting development in Texas captured the attention of a few housing industry observers. Toyota, the world's largest automobile manufacturer, set up a plant to construct about 50 modular homes for workers at a new automobile production facility in San Antonio. Although they have not indicated any plans to expand production beyond these homes, Toyota's presence is significant because advanced factory production systems employed at their plants in Japan would have potentially strong competitive advantages over U.S. production housing. First, visitors to Toyota's production facilities report their in-plant controlled environment production processes are much more efficient with far less waste while providing substantially better quality, trim, and attention to detail.[1] Second, the average construction time from order to completion is about 45 days compared to about 90 days for U.S. production housing. Third, American consumers have associated Toyota with powerful brand attributes such as quality, efficiency, and reliability prior to recent setbacks. There is no similar brand advantage associated with any large national U.S. builder. Finally, Toyota provides a 60-year warranty for defects compared to only 1 year for U.S. builders. That's a staggering difference in willingness to back up product performance.

These impressive advantages aside, the Toyota plant in Texas is not yet a serious blip on the U.S. housing industry radar screen. The U.S. housing market feels there is no threat from foreign competition. Meanwhile, virtually every other industry struggles to remain competitive in an increasingly global marketplace on the basis of cost, quality, and performance. Over time, many U.S. industries have succumbed or substantially contracted. Many other industries retain only an assembly function, importing most major components. The result is a growing list of products once dominated by U.S. manufacturers that have shifted to production overseas (clothes,

[1] Emanuel Levy of the Systems Building Research Alliance, personal communication, November 2009. Levy has led tours of European and Japanese housing plants for members of the U.S. housing industry.

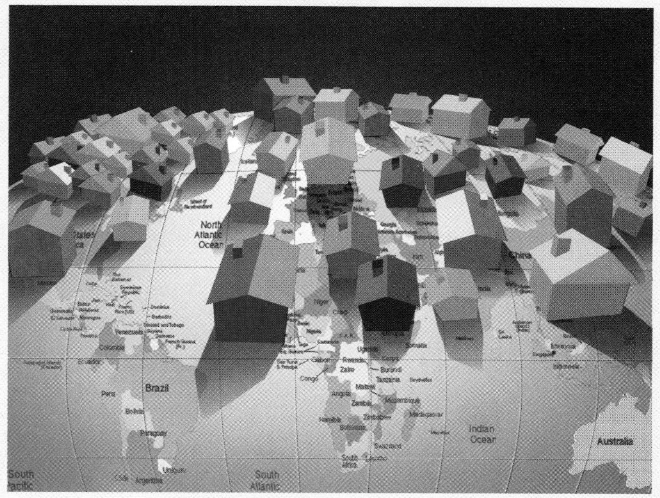

televisions, computers, phones, furniture, and the list goes on). Whole cities and regions have collapsed because of lost industries, including textile centers in New England and furniture centers in the mid-Atlantic, leaving numerous ghost towns in their wake. Many other industries are losing ground as they experience increasing losses to foreign competitors. Most baby boomers can remember when it was unusual to see imported cars. In 1965, U.S. automobile manufacturers accounted for more than 90 percent of domestic sales; they accounted for less than 45 percent of domestic sales in 2009.[2] Similarly, large commercial and small aircraft sales are losing substantial market share to global competition. Other industries that have been more resistant to global competition are now showing signs of losing their dominance in domestic sales. Asian and European manufacturers are making inroads in the large household appliances market, and foreign competition has begun to infiltrate the heating and cooling equipment market, which had been dominated almost exclusively by U.S. manufacturers.

As a result of these developments, U.S. manufacturing infrastructure appears to be dissolving before our eyes. Many experts express concern that the country's eroding manufacturing capabilities are undermining national security. U.S. dependence on foreign sources for major products makes us much more vulnerable should global conflict restrict world trade.

This brings us back to housing. As other American industries struggle to remain competitive with foreign companies, the importance of a strong U.S. housing industry to the U.S. economy increases. I believe the housing industry can effectively respond to the current

---

[2] Burton Marcus, "U.S. Automobile Manufacturers and Dealers Decline Precedes Revitalization," www.Marcumllp.com, December 2010.

financial crisis with the retooling recommendations included in this book, but a bigger question to consider is whether it will have a permanent reprieve from global competition. If not, can U.S. builders compete effectively?

A forthcoming book in the U.S. Housing Industry Review series, *How the World Builds Homes: Does U.S. Housing Stack Up?* will examine the competitive position of the U.S. housing industry. It will compare architectural styles, layouts, size, building materials, production processes, quality, and performance for homes constructed in the United States and in other industrialized countries. That little housing plant in Texas could be another warning shot.

Index

D

E

I

J

K

L

M

N

O

P

Q

R

S

T

X, Y, Z